Interregional Air Pollution Modelling

The State of the Art

NATO • Challenges of Modern Society

A series of edited volumes comprising multifaceted studies of contemporary problems facing our society, assembled in cooperation with NATO Committee on the Challenges of Modern Society.

Interregional Air Pollution Modelling

The State of the Art

Edited by

S. Zwerver

Ministry of Housing, Physical Planning, and Environment
Leidschendam, The Netherlands

and

J. van Ham

TNO Study and Information Centre for Environmental Research
Delft, The Netherlands

Published in cooperation with
NATO Committee on the Challenges of Modern Society

PLENUM PRESS • NEW YORK AND LONDON

Library of Congress Cataloging in Publication Data

Main entry under title:

Interregional air pollution modelling.

(NATO challenges of modern societies; v. 9)
"This volume comprises the final report of Panel 2 on Air Quality Modelling of the NATO CCMS Pilot Study on Air Pollution Control Strategies and Impact Modelling"—CIP copr. p.
Bibliography: p.
Includes index.
1. Air—Pollution—Meteorological aspects—Addresses, essays, lectures. 2. Air—Pollution—Mathematical models—Addresses, essays, lectures. 3. Atmospheric chemistry—Addresses, essays, lectures. I. Zwerver, S. II. Ham, J. van. III. North Atlantic Treaty Organization. Committee on the Challenges of Modern Society. IV. NATO CCMS Pilot Study on Air Pollution Control Strategies and Impact Modelling. Panel 2 on Air Quality Modelling. V. Series.
QC882.I59 1985 628.5'3'0724 85-19435
ISBN 0-306-42108-9

This volume comprises the Final Report of Panel 2 on Air Quality Modelling of the NATO CCMS Pilot Study on Air Pollution Control Strategies and Impact Modelling

A Division of Plenum Publishing Corporation
233 Spring Street, New York, N.Y. 10013

Printed in the United States of America

Members of Panel 2 on Air Quality Modelling of the NATO CCMS Pilot Study on Air Pollution Control Strategies and Impact Modelling

F. Axenfeld
Dornier System GmbH, Abt. NTUL
Friedrichshafen, Federal Republic of Germany

A. Berger
Institute of Astronomy and Geophysics
Université Catholique de Louvain
Louvain-la-Neuve, Belgium

M. C. Bovenkerk
Ministry of Housing, Physical Planning and Environment
Leidschendam, The Netherlands

P. J. H. Builtjes
TNO Division Technology for Society
Apeldoorn, The Netherlands

G. Clerici
Osservatorio Meteorologico di Brera
Milano, Italy

H. van Dop, *vice-chairman*
Royal Netherlands Meteorological Institute
De Bilt, The Netherlands

N. D. van Egmond
Institute for Public Health and Environmental Hygiene
Bilthoven, The Netherlands

K. E. Grønskei
Institute for Air Research
Lillestrøm, Norway

J. van Ham, *secretary*
TNO Study and Information Centre for Environmental Research
Delft, The Netherlands

W. Klug
Institut für Meteorologie TH Darmstadt
Federal Republic of Germany

C. Ludwig
Umweltbundesamt
Berlin, Federal Republic of Germany

J. C. Oppeneau
Ministère de l'Environnement et du Cadre de Vie
Service d'Affaires Scientifiques
Neuilly/Seine, France

L. P. Prahm
Air Pollution Lab.
Research Establisment Risø
Roskilde, Denmark

A. E. Saab
Electricité de France,
Direction des Etudes et Recherches
Chatou, France

F. Schiermeier, *NOAA (MD-80)*
Chief, Terrain Effects Branch
Meteorology and Assessment Division
Environmental Sciences Research Laboratory EPA
Research Triangle Park, USA

R. Stern
Freie Universität Berlin
Fachrichtung Meteorologie
Berlin, Federal Republic of Germany

H. E. Turner
Atmospheric Environment Service
Downsview, Ontario, Canada

M. L. Williams
Warren Spring Laboratory
Stevenage, Herts, United Kingdom

C. de Wispelaere
Programmatie van het Wetenschapsbeleid
Brussels, Belgium

S. Zwerver, *chairman*
Ministry of Housing, Physical Planning, and Environment
Leidschendam, The Netherlands

FOREWORD

The North Atlantic Treaty Organization (NATO) established the "Committee on the Challenges of Modern Society" (CCMS) at the November 1969 meeting of the North Atlantic Council.

The CCMS was charged with developing meaningful environmental and social programs that complement other international programs, and with showing leadership, first, in solution of existing problems and, second, in development of long-range goals for environmental protection in the NATO sphere of influence and in other countries as well. A first Pilot Study on Air Pollution was initiated by the CCMS at its inaugural meeting in December 1969. It resulted in documents about the definition of criteria for the effects of air pollutants as well as the development of assessment methods for air quality in urban areas.

A second Air Pollution Pilot Study (1975-1979) worked out the basics for setting up assessment methods for emissions inventories, techniques for the practical application of meteorological diffusion models as well as the development of guidelines for an Air Quality Management System (AQMS). Within this second Air Pollution Pilot Study attention to modelling concentrated on the Gaussian Plume Model.

A third Pilot Study on Air Pollution Control Strategies and Impact Modelling then was initiated in 1979 and started in 1980. This study was organized in three panels, respectively working on the Emission of Heavy Metals, on Interregional Modelling and on Impact Assessment. The pilot country for this study was the Federal Republic of Germany; co-pilots were the Netherlands for Panel 2 and the United States of America for Panel 3.

The results of the pilot study have been documented in a number of reports to the CCMS. They have been listed in the summary report that was submitted to the CCMS in the fall of 1984 (NATO-CCMS Document 150).

The rather serious damage to aquatic and terrestrial ecosystems, especially forests, emerging more and more during the course of the Pilot Study, prompted an increasing interest in interregional modelling of air pollution. In view of this development the pilot country made the suggestion to the Committee to further the dissemination of the body of knowledge that was collected during the past 5 years on the latter subject.

I have no doubt that the present volume in the series on Air Quality Modelling of Plenum Publishing Company will contribute to the solution of the serious environmental problems that now manifest themselves in large parts of Europe and North America.

Erich Weber
Chairman, Pilot Study

CONTENTS

SUMMARY

This document deals with the state of the art of interregional modelling. It is one of the results of the NATO-CCMS pilot study "Air Pollution Control Strategies and Impact Modelling" and was produced under the aegis of the Panel on Air Quality Modelling. The Panel selected the subject of interregional modelling and acted as a steering group during the study. The Netherlands, which was a co-pilot for the study, chaired the Panel. Ten countries participated.

Interregional modelling refers to the mathematical description of transport of air pollution over distances between 50 and 5000 km, covering areas between the size of countries or provinces and that of continents. The interregional scale may be subdivided into the mesoscale (50-500 km) and the synoptic scale (500-5000 km).

The choice of interregional modelling may be regarded as a logical extension of the activities of the earlier pilot study: "Air Pollution Assessment Methodology and Modelling". In that study the Gaussian plume model, which is considered to be valid for distances up to 25 km, was treated at length, among other things by an intercomparison of models for the Frankfurt area.

The average lifetime of common air pollutants like sulphur dioxide, nitrogen oxides, hydrocarbons and ozone is in the order of one or more days. Transport as well as chemical transformations which occur during their atmospheric residence time preclude a simple relationship between emissions and atmospheric concentrations.

An understanding of this relationship is vitally important for the development of control strategies, a view that is reflected in operational Air Quality Management Systems. The only way to establish relationships between emissions and atmospheric concentrations and, as a derivative, between emissions and deposition is by means of interregional models.

INVENTORY OF MODELS

Within the range indicated, 32 models were identified in 1981. In the present report 35 models are reviewed.

The essence of modelling is the approximation of reality through a mathematical analogon. In practice it is found that more than one approach can be chosen; the main categories are:

- analytical models which make use of the statistics of wind, diffusion and precipitation;
- eulerian models which describe all relevant properties of the atmosphere at fixed points in space;
- lagrangian or trajectory models which describe the processes in a coordinate system moving along with the wind;
- hybrid models which unite features of both eulerian and lagrangian models.

Models may, furthermore, differ in their goal and type of application, for example with respect to class of pollutant (inert or chemically reactive) or averaging time (episodic or long-term).

A more detailed review of the different types of models is presented in Chapter 2, where the various elements of a model are considered. Successively, atmospheric transport, requirements for different pollutants, deposition and chemical processes, input requirements (e.g. emissions, wind, mixing height, precipitation and air-quality data) and numerical methods are touched upon.
This type of analysis is then applied to each of the 35 identified models and a characterization of each individual model is presented. A summary of this analysis is presented in Table 1.

TRANSFORMATION AND REMOVAL PROCESSES

As the distance between source area and receptor area increases, the transformation and deposition processes will have more impact on the concentrations in the receptor area. For this reason interregional models are very sensitive to errors in the parametrization of these processes.

In Chapter 3 the state of the art with respect to deposition and chemical transformation is reviewed.

After a discussion of the deposition phenomenon and the methods of measurement for dry and wet deposition, the available data on the deposition velocities of sulphur dioxide, nitrogen oxides and sulphate are reviewed and their validity is assessed. For SO_2 and sulphate, consensus values can be derived; for nitrogen oxides, the body of information is insufficient to do so.

Table 1. Summary of models according to the classification given in section III. The symbols are explained in footnotes.

Model	Pollutant[1)]	Scale[2)]	Type[3)]	mathematical framework[4)]	averaging time[5)]	costs of operation[6)]
ADOM	3	s	c/d	e	e	m
AES	1	s	c/d	l	l	l
AIRSOX	2	s	c/d	h	l	l
ARL	1	s/m	c/d	h	e	l
ASTRAP	2	s	c/d	h	l	l
CEGB (SS)	2	m	c	e	e	l
CEGB (TD)	2	s	c/d	l	e	l
CIT	3	m	c	e	e	h
EPA	2	s	c/d	l	e	l
EPRI/ERT	3	s	c/d	e	e	h
EURMAP/ENAMAP	2	s	c/d	l	l	m
Fisher	2	s	d	a	l	l
Gillani	2	m	c/d	l	e	l
Hov	3	s	c/d	l	e	l
KNMI	3	m	c/d	e	e	h
Liraq	3	m	c	e	e	h
Mesoplume	2	s/m	c/d	l	e	l
Mesopuff	2	s/m	c/d	l	e	l
Mesogridd	2	s	c/d	e	e	m
Mesos	4	s/m	dos	l	e/l	l
OECD (ECE)	2	s	c/d	l	e/l	m
OME	1	s	c/d	l	l	l
Pseudo Spectral	2	s	c/d	e	e/l	l
RAPT	2	s	c/d	l	l	l
RCOM	2	s	c/d	a	l	l
RIV	1	m	c/d	l/e	e	l/m
SAI	3	m	c	e	e	h
SAI/EPA	2	s	c/d	e	e	m
STEM	3	s	c	e	e	h
Tald	4	s	dos	l	l	l
TDMB	1	m	c	e	e	m
Teknetron	2	s	c/d	e	e	l
TNO	2	m	c	a	l	l
Veltischeva	2	s	c/d	e	e	m
2 BPUFF	4	s/m	c	l	e	m

1) 1 = inert pollutants, 2 = chemically reactive pollutants modelled with conversion rates or decay constants, 3 = various pollutants including (photo)chemical reactions,
4 = radioactive pollutants

2) s = synoptic, m = mesoscale

3) c = determines ambient concentrations, d = deposition oriented, dos = dosage oriented

4) a = analytic, e = eulerian, h = hybrid, l = lagrangian

5) e = episodic, l = long term

6) l = low, m = moderate, h = high

Subsequently, the more important chemical transformation processes, such as reaction with the hydroxyl radical, are selected and compared with the data from field measurements. For sulphur dioxide and nitrogen oxides, effective first-order conversion rate constants are recommended. They appear to be seasonally dependent (for NO_x day-night variation should also be taken into account).

In addition, the information available on the removal of SO_2 and NO_x from plumes is reviewed.

INTERCOMPARISON OF MODELS

Chapter 4 describes the intercomparison of four interregional models when applied to a common data base. The design of this intercomparison project was detailed in Panel discussions and at an international expert workshop in Friedrichshafen. It was not considered very useful to make a comparison of models that are designed for different goals. Also, model input and output data should not differ too much. The intercomparison was directed towards episodic models. Other criteria for selection were:

- spatial resolution of 10-25 km;
- definite resolution in the vertical/models with two or more layers;
- removal by chemical processes should be accounted for.

From the 32 interregional models that had been identified in 1981, 10 passed this selection. On request, 4 of them were made available for the intercomparison study. These were the KNMI mesoscale transport model, the RIV puff model, an enlarged version of the SAI urban air-shed model (SAI-TNO) and the THD model. The Technical University of Darmstadt was chosen as the centre for the study.

The intercomparison was divided into three phases. A modelling area of dimensions 350x500 km^2 covering the Netherlands, Belgium, the Western part of the Federal Republic of Germany and a small part of Northern France was used throughout the study.

In Phase I numerical tests of the models were carried out against the analytical solutions of some hypothetical dispersion experiments; chemical and deposition processes were absent in these calculations. These tests showed that the numerical accuracy of the models is adequate for all practical applications.

In Phases II and III data bases for three actual meteorological episodes served as input for the model. In Phase II there is only one (hypothetical) continuous source of fixed position and strength. In Phase III computations are carried out based on the emission data base for SO_2 in the modelling area.

In the latter two phases differences between the model output are found to be much more pronounced than in Phase I. This leads to the conclusion that these differences are caused by the - sometimes completely - different processing of the meteorological input data to the model core. This concerns the wind field and the mixing height, in particular. The benefit of using empirical data on mixing height during the modelled episodes appeared to be substantial.

An important recommendation from the intercomparison study, therefore, is to improve methods for the derivation of wind field and vertical structure from routine meteorological data.

Reviewing the different episodes, it appears that all models have their weak and strong points: in a given meteorological situation one will be better than the others, but will prove inferior on another occasion. Generally speaking, agreement between the models is better for calculated concentrations than for calculated dry and wet deposition. Improvement of the deposition element in the tested models is recommended.

For a comparison between modelling results and measured ambient concentrations, initial concentrations and fluxes across the boundaries of the modelling area should have been taken into account. Notwithstanding, the correlation between the models and reality is reasonably high most of the time.

CONCLUSIONS

The intercomparison study shows distinctly that interregional, episodic modelling on the mesoscale (500 km) and with substantial spatial resolution has reached the state of applicability for sulphur dioxide in appropriate areas. Calculated concentrations yielded by models designed for the same goals do not differ substantially within present expert knowledge. At present, calculation of deposition deserves further development.

For sound application, however, an adequate emission data base and information on initial ambient concentrations and fluxes across the boundaries of the modelling area are required.

Projects on the intercomparison of models will certainly induce improvement of the state of the art of modelling, as has the present study.

CHAPTER 1

INTRODUCTION

S. Zwerver* and J. van Ham**

* Ministry of Housing, Physical Planning
and Environment, Leidschendam
The Netherlands

** TNO Study and Information Centre
for Environmental Research
Delft, The Netherlands

In 1980 the Committee on the Challenges of Modern Society decided to take up a new Pilot Study in the field of air pollution under the title of Air Pollution Control Strategies and Impact Modelling. The Pilot Study, under the pilotship of the Federal Republic of Germany, was chaired by Dr. Erich Weber of the Bundesministerium des Innern, Bonn. Mrs. Christa Morawa, Umweltbundesamt, Berlin, acted as the secretariat. The pilot study was divided into three Panels:

Panel 1 on Emission of Heavy Metals
Panel 2 on Air Quality Modelling
Panel 3 on Environmental Impact

For Panel 2 The Netherlands took responsibility as a co-pilot. Representatives from ten countries participated in the Panel. At the first meeting of the Working Group for the Pilot Study in April 1980 the subject of Interregional Modelling was selected as being most relevant, and suitable for study within an international framework. It had been observed that modellers in several countries in Europe and North America were working on the subject. The interest for the subject sprang from the general awareness that air pollution does not stop at national boundaries but may travel over distances of several thousand kilometres.

The subject of interregional modelling is an extension of work performed in the preceding Pilot Study on Air Pollution Assessment Methodology and Modelling. In the latter study the Gaussian plume model was treated and recommended for application in national Air Quality Management Systems. The Gaussian plume model is suitable for the description of dispersion over distances not exceeding 25 km.

For the control of those air pollution problems which play a role at greater distances, up to the continental scale, different approaches are required. The models used should cover areas of the size of the transport phenomena. Moreover, since control strategies should be developed in international consultation it is important to base these strategies on models which are acceptable on an international level. It was considered that a review of available models in the range of 50-5000 km might improve the acceptability of many of these models.

The wide range mentioned above reflects the different approaches that have been chosen by scientists in different countries:

- the long-range transport model, which covers distances of several thousand kilometres.

This approach takes in the scale of the transport phenomena in the atmosphere. However, data bases tend to be rather incomplete for areas of continental dimensions. Moreover, attention to detail has to be abandoned.

- the interregional model, which is applied to areas for which adequate data bases exist.

In these models more detail can be retained, which is important for validation of such a model. Also, when the area for which a data base exists increases, the model can be subsequently enlarged.

Atmospheric models simulate the processes that start with emissions of pollutants and that result in ambient concentrations of pollutants; more specifically, a model should deal with dispersion, transformation and removal, taking into account that these processes occur simultaneously. For a comparison of modelling results with measured concentrations, it is essential to consider the initial concentrations within the modelling area and the fluxes from outside the area (see diagram).

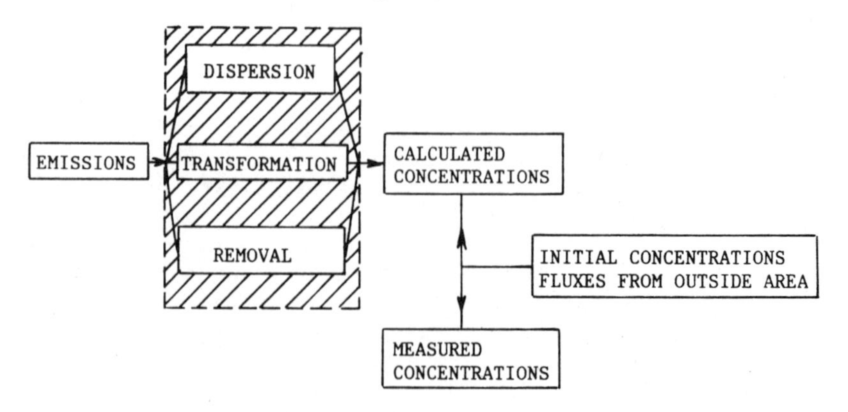

The Panel decided to undertake a threefold effort in the field of interregional modelling:

- to make an inventory of available interregional models and to review these with respect to their main characteristics and parameterization;
- to review recent knowledge on chemical transformations and removal processes, in acknowledgement of the fact that these processes become progressively more important as the size of the modelling area is increased;
- to make a more detailed intercomparison between a sample of the available models.

The present document reports the results of the Panel studies which relate to the shaded area in the diagram.

In Chapter 2, which was prepared for the Panel by K.D. van den Hout and H. van Dop, the review of interregional models is presented. An earlier version of this chapter was published in 1981 as NATO-CCMS document No. 126. It was updated during June 1984.

In Chapter 3, which was prepared for the Panel by R.M. van Aalst and H.S.M.A. Diederen, transformation and removal processes are reviewed. An earlier version of this chapter was published as NATO-CCMS document No. 127, also in 1981. It was updated during February 1985.

In Chapter 4, an intercomparison of four different interregional models under various conditions is reported.
The intercomparison project was carried out by a team of scientists from the Federal Republic of Germany and the Netherlands and was headed by W. Klug (Technical University of Darmstadt).
This study was approved by the Panel in March 1984.

CHAPTER 2

INTERREGIONAL MODELLING

K.D. van den Hout* and H. van Dop**

* TNO Environmental Research Institute
Delft, The Netherlands
** Royal Netherlands Meteorological Institute, De Bilt
The Netherlands

SUMMARY

A review is given of (in principle) operational models describing the transport, diffusion and chemical transformation of air pollutants to a horizontal scale varying from 50 to 5000 km. In the first part of the report, the general features of interregional transport models and the various formulations used to describe the atmospheric processes are briefly reviewed. The following aspects are discussed separately: spatial scale, atmospheric transport, air pollutant components, chemical processes, deposition, input requirements and numerical methods. The second part of the report contains a description of 32 models. For each model, it is indicated how the above-mentioned aspects are treated.

I. NOTATION

C = concentration ($\mu g.m^{-3}$)

H = mixing height (m)

$K_{x,y}, K_H$ = horizontal eddy diffusivity ($m^2 s^{-1}$)

K_z = vertical eddy diffusivity ($m^2 s^{-1}$)

L = Obukhov length (m)

P = Pasquill stability class

R = precipitation rate (mm/hr)

q = wet deposition rate (hr^{-1})

u, v, w = horizontal (u, v) and vertical (w) component of the wind velocity vector (ms^{-1})

u_* = friction velocity (ms^{-1})

V_d = deposition velocity (ms^{-1})

β = $SO_2 \rightarrow$ sulphate conversion rate (hr^{-1})

Λ = radioactive decay constant (hr^{-1})

σ_y = horizontal dispersion coefficient (m)

σ_z = vertical dispersion coefficient (m)

II. INTRODUCTION

In this document, a review is given of (in principal) operational models which describe the transport, diffusion and chemical transformation of air pollutants, to a horizontal scale varying from 50 to 5000 km. With this spatial scale corresponds a time scale varying from a few hours to a few days. At these scales, chemical transformation processes, decay and deposition processes are of equal importance compared with transport and diffusion processes. Models to a smaller scale than 50 km using the characteristic approximations corresponding to that scale are excluded from this report. Also global models where the vertical distribution of pollutants in the whole troposhere has to be taken into account are excluded.

First, a brief review will be given of the general features of interregional transport models and of the various formulations used to describe the physical and chemical processes (sections III - VIII). In the second part, the properties of each model separately will be described in tabulated form (section IX). It is not attempted to characterize the applicability or accuracy of the various models. These matters will be dealt with in Chapter 4 of this volume.

III. GENERAL FEATURES OF INTERREGIONAL TRANSPORT MODELS

Interregional transport models can be classified according to their objective and their spatial and temporal scales. Further important distinctions are their mathematical approach and complexity. In this section we shall give a brief description of these aspects.

Goal

The objectives of a model are strongly related to the kind of pollutant. In the most general case the basic dispersion model should have the capability to predict *concentrations*; usually, these are averages over a given volume and period of time. When the major effects of air pollution are due to concentrations in air it suffices to calculate ground level concentrations (e.g. photochemical air pollution). In other cases air concentrations are related to other quantities such as (acid) *deposition* or (radioactive) *dosage* models, respectively. The dosage models are mainly used in studies of releases of radioactive material.

Spatial Scale

In the introduction it has already been stated that only models are considered which operate to a scale from 50 to 5000 km; the small scale (mostly Gaussian plume type) models, on the one hand and the global models, on the other are thus excluded. It is useful to make a second subdivision in scale based on a comparison with the density of the global radiosonde stations network. This density greatly varies over oceanic and continental regions. A typical interstation distance in more populated areas may be taken to be 50 km. As a consequence, meteorological input data for models to a scale of 50 - 500 km are usually derived from the more dense network of surface observation stations, while in models to a scale exceeding 500 km in general radio sonde data are applied. The models will be referred to as *mesoscale* models and *synoptic scale* models, respectively. These are rather arbitrary definitions, and the authors realize that in literature many other "typical distances" are assigned to, e.g., "meso scale." It should be noted that with increasing spatial scales, the observation frequency decreases, so that meteorological data become sparse in time.

In Table III.1, a review is given of the scales defined in the context of this document and their specific meteorological input requirements.

Averaging Time

Another subdivision can be made with regard to averaging time. If the model is designed to predict (order of) hourly average concentrations during a limited amount of time, say a day or two, the model has an *episodic* nature. In episodic models for scales larger than 50 km, the variation of the meteorological conditions in time and space is taken into account. In *long term* models (with averaging times in the order of a year), the description is often simplified: calculations are not carried out for every hour, in some cases the meteorological variations in time and space are further parametrized.

Table III.1 A classification of models according to their spatial scales.

Model type	Range km	Typical model configuration			
		observations used	wind fields	diffusion	data frequency (hr^{-1})
micro scale	0-50	surface synoptic data or special measurements	uniform or varying in time and space	homogeneous (dispersion coefficients) or non-homogeneous (eddy diffusivity concept)	~ 1
meso scale	50-500	surface synoptic data and radio sonde data	varying in time and space	non-homogeneous (eddy diffusivity concept)	1 - 0.3
synoptic scale	500-5000	radiosonde data	varying in time and space	non-homogeneous (eddy diffusivity concept) or instantaneous mixing	0.3 - 0.1

Mathematical Framework

Finally, models can be divided according to their mathematical framework. *Eulerian* models describe all the relevant physical properties at fixed points in space, while *Lagrangian* (or trajectory) models describe the processes in a co-ordinate system moving along with the wind. Some models have both Langrangian and Eulerian features and are therefore called *hybrid* models. So-called *analytical* models do not perform a numerical simulation of the atmospheric processes, but use analytical expressions based on statistics of wind, diffusion and precipitation.

From this and the following chapters, it is obvious that the degree of sophistication with which aspects such as emission, transport and chemistry are treated, varies considerably amongst the models (Eliassen, 1980). It is important to note here that the highest degree of sophistication is not necessarily its optimum degree. The design of the models reflects a compromise between conflicting requirements regarding physical/chemical realism and the costs of operation (which are mainly determined by input preparation and computing time).

IV. ATMOSPHERIC TRANSPORT

Most of the models use a deterministic description of the atmospheric transport; the general starting point is the well-known transport diffusion equation:

$$\partial c/\partial t + u\partial c/\partial x + v\partial c/\partial y + w\partial c/\partial z =$$

$$\partial/\partial x\ (K_x\ \partial c/\partial x) + \partial/\partial y\ (K_y\ \partial c/\partial y) + \partial/\partial z\ (K_z\ \partial c/\partial z). \qquad \text{(IV.1)}$$

In this equation, the turbulent dispersion is represented by the eddy diffusion coefficients, K_x, K_y and K_z. Several models use this equation directly (Carmichael and Peters, CIT, SAI model). In most cases simplifying assumptions are made:

- Often the vertical velocity, w, is so small that it can be neglected in practical applications. Consequently, $\partial u/\partial x + \partial v/\partial y$ should approximately be zero. When observed (interpolated) wind fields are used this condition should be fulfilled in order to conserve pollutant mass. Also numerical problems may rise by application of divergent wind fields.
- The approximation that horizontal turbulent diffusion can be neglected with respect to horizontal transport can be justified in some cases. The result is the equation:

$$\partial c/\partial t + u\partial c/\partial x + v\partial c/\partial y = \partial/\partial z\ (K_z\ \partial c/\partial z). \quad \text{(IV.2)}$$

This equation can be used for a mixed Lagrangian-Eulerian (hybrid) formulation. In fact, eq. (IV.2) can be divided into two equations:

$$\partial c/\partial t + u\partial c/\partial x + v\partial c/\partial y = 0, \text{ and} \quad \text{(IV.3)}$$

$$\partial c/\partial t = \partial/\partial z\ K_z\ (\partial c/\partial z). \quad \text{(IV.4)}$$

Eq. (IV.3) describes the advection of pollutant only and may be solved by determination of the (Lagrangian) trajectories. If vertical wind shear can be neglected ($\partial u/\partial z = \partial v/\partial z = 0$), the transport can be represented as a column of air which travels along trajectories, where during its travel vertical diffusion takes place according to the (Eulerian) diffusion equation (eq. IV.4). The eddy diffusivity, K_z, is then continuously adapted to the circumstances along the trajectory (ARL, BNL, Bolin and Persson, CEGB, Gillani, PNL model). Several models use empirical (Gaussian) dispersion coefficients for the horizontal and vertical diffusion, while retaining the description of advection according to eq. (IV.3) (Mesopuff, Mesos, Tald model).

- If also vertical diffusion is neglected ($K_z = 0$) or vertical homogeneity is assumed (well-mixed conditions) only eq. (IV.3) is retained. This equation is the basis of Lagrangian trajectory models (EURMAP, OECD model).
- Assuming stationary state conditions and simple expressions for the wind field and turbulence analytic solutions can be obtained (see e.g., Pasquill, 1974, Sutton, 1943). This approach cannot be adopted for the prediction of short-term concentration levels, since the deviations of the actual atmospheric transport from the assumed behaviour may be very large. For the calculation of long-term averaged values, however, such an analytical description in terms of statistics of wind and turbulence parameters is more accurate (Fisher, TNO model).

V. AIR POLLUTION COMPONENTS

SO_2 and Sulphate

Most of the interregional models deal with SO_2 and, usually, with its reaction product, sulphate (including H_2SO_4). Anthropogenic sulphur is emitted almost entirely as SO_2, only a small percentage is emitted as SO_3 or sulphate. In the industrialized regions, anthropogenic sulphur emissions dominate over the natural emissions. Because the natural emissions (partly H_2S and other compounds, which are converted to SO_2) are not well known and the natural contribution to the concentration is estimated to be relatively small in these areas (Granat et al., 1976), these emissions are not incorporated in most model applications.

NO_x and Photochemical Products

The anthropogenic contribution to the NO_x emission mainly consists of NO (about 85%) and NO_2. These gases are chemically more reactive than SO_2. The chemical transformation of NO to NO_2 depends on the O_3 concentration; the reverse transformation ($NO_2 \rightarrow NO$) is governed by a photolytic process which depends strongly on the solar radiation. For mesoscale calculations for NO_x, photochemical dispersion models are available at present (CIT, Liraq, SAI model). A chemical kinetic mechanism describes the interaction between several chemical components, of which O_3 is the primary component of interest. Because of the high costs of operation these models are applied for short-term calculations only.

"Acid Rain"

Recently damage on a large scale has been observed to forests and other ecosystems in Europe, the United States and Canada. There are strong indications that these damages are caused by "acid rain" or more accurately "acid deposition". The term "acid deposition" is preferred because it appears that approximately only 50% of the total deposition results from precipitation. The other half is removed from the air by "dry deposition". The major damage is thought to occur by a combination of an average high acid deposition and intermittently high photooxidant (ozone) concentrations. The major components in the acid deposition process are SO_2, NO_x and NH_3.

Radioactive Gases

Finally, a number of models describe the transport of radioactive releases (Mesos-, Tald-model). These models are intended for short-term calculations, usually for a relatively small cloud of material (puff). The radioactive decay of these compounds is specified through a first-order transformation rate; the compounds are chemically inert.

VI. CHEMICAL PROCESSES AND DEPOSITION

The distances over which pollutants can be transported are determined by the chemical transformation and wet and dry deposition processes. Since, in the case of sulphur oxides, the deposition rates of the secundary pollutants (sulphate aerosols) differ substantially from those of SO_2, a proper description of the transformation of SO_2 to sulphate is important, even if the calculations are aimed at long-range transport of total airborne sulphur. Under most weather conditions, SO_2 is removed primarily by deposition effects. Also the transformation to sulphate is an important removal mechanism with conversion rates up to a few per cent per hour. Of the deposition mechanisms, only the dry deposition is active in periods without rain with a typical removal rate of a few per cent per hour. During precipitation wet deposition dominates, but on the average the deposition due to precipitation is comparable to the dry deposition (with some notable exceptions like southern Norway) (e.g. Eliassen, 1978). The few models that can be used for NO, NO_2 and photochemical products are intended for application to warm and sunny episodes, and consequently do not contain a description of wet deposition.

A review of the experimental and theoretical research on the various loss mechanisms is given in Chapter 3 of this volume.

Chemical Processes

From the extensive research that has recently been executed, it has become clear that the chemical transformation of SO_2 to sulphate is a very complicated process. Homogeneous (gas phase) oxidation rates are comparable with heterogeneous (catalytic) oxidation rates; each of both processes depends in a complex way on the physical and chemical characteristics of the air mass. Since a suitable parametrization of all these processes is hardly feasible at present, almost all models contain a simple description of the transformation; in many cases a constant first-order transformation rate of 1% per hour is used. Some of the models use first-order transformation rates with daytime values different from those for nighttime (PNL-model), seasonally dependent values (TDMB model) or rates which are proportional to the actual solar radiation (Gillani model). In some models (Carmichael/Peters, CIT, ERT, Liraq, SAI model) a (photo)chemical reaction mechanism is included for the gasphase transformation of SO_2, which uses prespecified or calculated concentrations of several other components such as nitrogen oxides, hydrocarbons and radicals. The heterogeneous transformation in these models is accounted for by a first-order transformation rate, or is neglected.

The chemical transformation rates for other compounds (e.g., nitrogen oxides) are calculated in three photochemical models (CIT, Liraq, SAI model), which are primarily intended for simulations under weather conditions which prevail during photochemical air pollution episodes.

The transformation of radioactive compounds is not of a chemical nature. In the models describing the dispersion of radioactive releases the removal by radioactive decay is described by a decay constant, analoguously to the conversion rate description often employed for the $SO_2 \rightarrow$ sulphate conversion.

Dry Deposition

Dry deposition, i.e. removal of airborne pollutant by surface uptake, is included by using the concept of a deposition velocity, which is defined as the net downward flux divided by the concentration at a reference height. Often a reference height of 1 m is taken. As a result of the removal at the surface, a vertical concentration profile depending on the stability of the atmospheric boundary layer is established and hence the deposition velocity in principle is dependent on the reference height.

Not only is the deposition velocity dependent on atmospheric stability, but also the temperature and other meteorological quantities affect the deposition rate, in particular the uptake by plants. Differences in surface characteristics such as the surface material and roughness may further result in differences in effective deposition velocity of one order of magnitude or more. Most models use a constant value for the deposition velocity of a species which is assumed to be representative for the whole modelling region. The deposition velocity may be taken proportional to the friction velocity (e.g. the SO_4 deposition in the model of Carmichael and Peters), thus relating the vertical flux to the turbulent diffusion. Often more surface characteristics (roughness, evaporation, stability, wind speed) are included in the description of deposition (Carmichael and Peters, SAI, CIT and ERT models). In some low resolution models, initial deposition effects, associated with the strongly non-uniform vertical concentration profiles close to the sources, are separately taken into account. The OECD model assumes an initial SO_2 deposition in the grid cell of emission of 15%, and EURMAP-II takes the deposition velocity inversely proportional to an assumed puff height.

Wet Deposition

The importance of an accurate description of wet deposition is largely dependent on the purpose of the model calculations. Obviously, for a study of the acid rain problem, the model should contain more detailed mechanism than for a calculation of average atmospheric concentrations at relatively small distances from the source, where the loss due to precipitation is rather small. Several investigators have shown that during short episodes large amounts of sulphur can be deposited by precipitation, which may be of major importance for the understanding of the effects of acid rain (see, e.g., Smith, 1980).

At present, only a limited insight in the processes of washout (removal of pollutants by precipitation below the clouds) and rainout (in-cloud removal) exists. The models do not treat these two processes separately; either a description is given by one parameter (often a scavenging coefficient, which is the removal rate by precipitation), or the wet deposition is not taken into account at all. A complication arises from the limited time and space resolution of the precipitation data and of the model itself: precipitation data are routinely available as six hourly sums, and the distance between precipitation stations ranges from a few tens to hundreds of kilometres. Especially convective storms which may remove almost all SO_2, typically have a duration and horizontal extent which are too small to be resolved.

Several of the models incorporate the wet deposition rate as a first-order removal process which is uniform in time and space (EPA, TNO model). Most other models subject precipitation observations to some interpolation procedure and then take the wet deposition rate either dependent on the precipitation rate (ENAMAP, EURMAP, OECD, PNL model) or constant during precipitation (Bolin/Persson model). In the OECD model, atmospheric concentrations of SO_2 and sulphate were calculated using a constant wet removal rate during rain, while the wet deposition pattern for sulphur was determined using a linear relation between the concentration of airborne sulphate and the sulphur concentration in precipitation. In the Fisher model, the wet deposition is derived from typical values for the duration of wet and dry periods, which are allowed to vary in space. In the photochemical dispersion models which can be employed for the mesoscale transport of compounds like NO, NO_2, O_3 and SO_2, no wet deposition mechanism is included, because episodes of photochemical smog formation are usually dry.

Of the models describing the transport of radioactive material only the Mesos model contains a description of wet deposition.

VII. INPUT REQUIREMENTS

Usually the costs of operating the models are dominated by the gathering and preparation of the input data. An important characteristic of the models, therefore, are the amount and detail of the input data required. The input can be divided into two main categories: meteorological data (including terrain characteristics) and air pollution data (emissions and air quality data).

Meteorological Data

The differences in meteorological data requirements of the various models are substantial.

Wind

All models require information on the wind. Some long-term models are based on statistics of meteorological data of the wind speed and direction at a surface station at one or more locations that are considered representative (Fisher, TNO model). The other models, especially when directed to short-term calculations, use wind observations of the period considered. In some of these, only wind data measured at the 850-mbar level (about 1500 m high), which are routinely available every six hours from sonde measures, are used (ENAMAP, EURMAP, OECD, Pseudo Spectral model). For other models both 850-mbar and surface wind observations are required (EURMAP II, Tald model). Several models require vertical wind profiles as input, directly measured or derived from upper air wind observations (BNL, Mesogrid, Mesoplume, Meso puff, PNL, Teknetron model) or from upper air, tower and surface wind observations (Carmichael/Peters, CIT, EPA, SAI, Veltisheva model). The Bolin/Persson and Mesos model derive geostrophic winds from atmospheric pressure data. One model requires, instead of wind data, air trajectory data as input (Gillani model).

If the wind field used is not divergence-free, los or gain of pollutant mass may occur. Also numerical problems may arise when a non-divergent wind field is used: in a region of strong convergence very sharp gradients in pollutant concentration may occur, which is likely to result in large numerical errors. However, procedures have been developed to transform a non-divergent wind field into a divergence-free field (Endlich, 1967, Sherman, 1978).

Mixing Height

Many of the models use constant or internally prescribed time variations of the (horizontally constant) mixing height. The statistical models (Fisher, TNO model) use mixing height classifications based on wind and stability. Several models for short-term calculations use hourly specifications or daily maximum and minimum values (ARL, Carmichael/Peters, CIT, EPRI/ERT, Gillani, Liraq, Mesos, Teknetron model).

Precipitation

Precipitation measurements, usually available as 6-hourly totals, are required in many models. In the statistical models which use average wet removal rates, precipitation data are not required. A few models neglect precipitation, or are not intended for use in cases with precipitation.

Other Meteorological Data

For the description of the wind profile, turbulent diffusion and chemical processes, some models need further input data on temperature (surface and upper air), humidity (surface and upper air), cloud amount, solar radiation (UV), stability class, evaporation rates and roughness.

Air Pollution Data

Emissions

Several models (e.g. all Lagrangian models with homogeneous mixing up to the mixed-layer height) use only a limited spatial resolution (horizontally of the order of 100 km, vertically one cell of about 1 km deep). For these models, the emissions are taken to be instantaneously mixed in grid cells of comparable dimensions. Information on the source height is of importance for the determination of the initial deposition (close to the source) and, furthermore, to allow sources to emit above the mixing height. For models with finer horizontal resolution (~ 10 km or less) where the spatial concentration pattern close to the source can be resolved, the source height and plume rise parameters for the major point sources are required. Smaller sources are considered as area sources at surface level, or are ignored. The temporal variation of emissions is not taken into account in some of the models; for other models, a fixed annual variation in emission is prescribed for long-term calculations, or, for short-term calculations, the emissions may be specified every hour. In synoptic scale calculations, the accuracy requirements for individual emissions are usually less important than for microscale calculations, because of the large amount of sources involved and the many uncertainties in other important parameters. However, because many sources in different countries or states with different regulations and different fuel usage are considered, estimates of even the total emissions on a low resolution grid are often unreliable.

Air Quality Data

Since all models include the transport of emissions only explicitly for the sources within the modelling region and the modelling period, the effect of emissions outside the region (background contribution) or previous to the period is either neglected or estimated from air quality measurements (or from results of other models). Since transport distances are typical for sulphur compounds in the range of 100 - 2000 km, the background sulphur contribution is neglected for model regions with dimensions of a few thousand km. The initial sulphur concentrations for the trajectory models are set at zero. For smaller scale models, an estimate of the background contribution is included (TNO model) or, for grid models for short term calculations, initial concentration fields and hourly values of boundary concentration fields are estimated (Carmichael and Peters, CIT, EPRI/ERT, Liraq, Mesogrid, SAI, SAI/EPA, TDMB). Models containing chemical reaction schemes require air quality specifications for several compounds.

VIII. NUMERICAL METHODS

In this section, the various numerical methods used in the models are briefly reviewed.

The most simple from a numerical point of view are the models using analytical solutions for the concentration field. These methods are numerically trivial and efficient; the derivation of the equations, however, involves many important approximations.

More realistic wind and turbulent fields can be included in the Lagrangian (trajectory) and Eulerian models (see sections III and V). Of these, the Lagrangian models allow a straightforward and rather efficient formulation of the atmospheric transport when vertical homogeneity of the wind is assumed.

The most complete description of transport, turbulence and chemistry can be given in an Eulerian model. A drawback of this approach is the computational effort needed to integrate the transport equations numerically. For a 3-dimensional grid with an appreciable number of grid points, this is a costly procedure. Designers of Eulerian models therefore pay considerable attention to the choice of the numerical technique.

In this section, we will give a review of numerical techniques used. Of these, the finite difference method is the most basic approach to all differential equations. In addition, a number of methods specially used for the solutions of the advection equation are mentioned: Galerkin method, the method of moments, the particle in cell method, and the Shasta method. The finite difference methods for the diffusion equation will be briefly discussed.

Finite Difference Methods

The essence of the method is that the differentials which occur in the transport equation are replaced by finite differences. For instance, in the one-dimensional advection equation (cf. eq. (IV.3))

$$\partial c/\partial t + u\, \partial c/\partial x = 0, \qquad \text{(VIII.1)}$$

where $c(x, t)$ is the concentration and $u(x, t)$ the wind field, the first derivative with respect to time, $\partial c/\partial t$, can be replaced by a forward difference:

$$\partial c/\partial t = (c^{n+1} - c^{n})/\Delta t + 0(\Delta t) \quad \text{(VIII.2)}$$

where $n + 1$ and n correspond to the concentration field at time $t+\Delta t$ and t, respectively. The second term in eq. (VIII.2) denotes higher derivatives containing powers of Δt of order one and higher. Correspondingly, the spatial derivative can be written as a central difference:

$$\partial c/\partial x = (c_{i+1} - c_{i-1})/2\Delta x + 0(\Delta x^{2}) \quad \text{(VIII.3)}$$

where the integers $i+1$ and $i-1$ correspond with the position $x+\Delta x$ and $x-\Delta x$, respectively. Rewriting eq. (VIII.1) using eqs. (VIII 2,3) yields

$$c_{i}^{n+1} = c_{i}^{n} - \lambda/2\ (c_{i+1}^{n} - c_{i-1}^{n}) + 0(\Delta t\Delta x^{2}). \quad \text{(VIII.4)}$$

The Courant number, λ, is $u\Delta t/\Delta x$. The accuracy of the scheme is determined by the magnitude of Δt and Δx. Moreover, the value of the Courant number determines whether the difference scheme is "stable". Stability implies that if the solutions of the original differential equation are bounded, the numerical solution is also bounded. In general, λ should be chosen smaller than some constant, which is determined by the choice of the difference scheme. (In the example (eq. VIII.4), no such value exists, so that this particular scheme is always unstable).

If the domain of integration and the initial and boundary conditions are specified, the set of linear equations (eq. VIII.4) can be solved. The above difference scheme is called explicit in time, because the value of c at time $t+\Delta t$ is computed from an expression which contains terms depending on times $t' < t+\Delta t$ only. The accuracy of this first-order difference scheme is only acceptable for very small increments in time and space. Improvement of the accuracy can be obtained in general by choosing higher-order approximations. Different types of differential equations require different finite difference approaches (Richtmeyer and Morton, 1967), which constitute a separate field in applied mathematical research.

Galerkin Methods

A common characteristic of these methods is that the concentration, c (in general the dependent variable), is expanded in a series of (orthogonal) functions, $g_k(x)$, according to:

$$c(x,t) = \sum_{k=1}^{N} c_k(t)\, g_k(x). \qquad \text{(VIII.5)}$$

In the finite element method (Strang and Fix, 1973), mostly simple functions are used as the basis functions $g_k(x)$. In general, finite element methods lead to finite difference methods. An advantage of the method is that it is particularly suited for problems with irregularly shaped model regions and boundary conditions.

A special case of the Galerkin method is the Fourier expansion ($g_k(x) = \exp ikx$). The method is called pseudo-spectral, if it is used in the spatial integration only. Time integration is usually done by a third-order difference scheme (Prahm and Berkowicz, 1977). A drawback of the method is that the mathematical treatment requires that the concentration distribution should be continued periodically outside the integration region. As a consequence, the boundary condition at the inflow side of the region is determined by the boundary condition at the outflow side, which is physically unrealistic. Therefore, special measures should be taken near the boundary of the region. Nevertheless, the pseudo-spectral method seems a very suitable method for the solution of the advection equation (De Haan, 1980). The easy access to fast Fourier transform numerical techniques makes this method very efficient.

Method of Moments

Instead of finite differencing eq. (VIII.1) directly, the moments of the concentration distribution c(x,t) within each grid cell are introduced as follows:

$$\bar{c}_m = 1/\Delta x \int_{x_m-\frac{1}{2}\Delta x}^{x_m+\frac{1}{2}\Delta x} c(x,t)dx,$$

$$\bar{x}_m = 1/\Delta x \int_{x_m-\frac{1}{2}\Delta x}^{x_m+\frac{1}{2}\Delta x} c(x,t)x\,dx/\bar{c}_m, \text{ and} \qquad \text{(VIII.6)}$$

$$\sigma_m^2 = 1/\Delta x \int_{x_m-\frac{1}{2}\Delta x}^{x_m+\frac{1}{2}\Delta x} c(x,t)(x-\bar{x}_m)^2 \, dx/\bar{c}_m.$$

The average concentration, the center of mass and the variance of the concentration distribution within each grid cell are denoted by $\bar{c}_m$, $\bar{x}_m$ and σ_m^2 respectively. By manipulation of the transport equation, three equations can be obtained for the (first) three moments. These equations are solved with a simple forward finite difference method. In this way a solution is obtained with an accuracy which would not have been achieved by applying the (same) finite difference scheme to the original transport equation (Egan and Mahoney, 1972).

The Particle in Cell Method

This method is essentially based on a Lagrangian technique. For a source, particles are continuously emitted. The particle trajectories are computed according to:

$$X_i(t+\Delta t) = X_i(t) + \Delta t \; V_i(t), \qquad \text{(VIII.7)}$$

where X_i and V_i refer to the position and velocity of the ith particle, respectively. The amount of particles within each grid cell corresponds with the total pollutant mass. Diffusion is included by means of the "turbulent flux velocity" concept (Sklarew et al., 1971, Lange 1978).

$$U_f = -K/c \; \partial c/\partial x, \qquad \text{(VIII.8)}$$

so that the final transport equation reads:

$$X_i(t+\Delta t) = X_i(t) + \Delta t \; [V + U_f]_i. \qquad \text{(VIII.9)}$$

(In eq. (VIII.9), U_f has to be evaluated at the position of the ith particle.) A drawback of the method is that for large grids and many sources a large amount of particles have to be released and tracked.

Shasta

This scheme has been developed in order to maintain sharp gradients in concentration distributions in transport models (Boris and Book, 1973). It is a flux-corrected-transport algorithm. In addition to the mass-conserving property of most conventional algorithms, the FCT algorithm strictly maintains the positivity of actual mass densities. This is possible because an FCT algorithm consists conceptually of two major stages, a transport stage followed by a corrective stage. Both stages are conservative and maintain positivity. Their interaction prevents the usual dispersively generated ripples near sharp gradients.

The last three methods are attempts to circumvent errors due to numerical diffusion, which can be significant in low-order finite difference schemes.

Besides the above-described schemes, there exist a variety of numerical schemes for the solution of the advection equation, of which we mention the "leap-frog" scheme (e.g. Mesinger and Arawaka, 1976), the Lax-Wendroff scheme (Richtmeyer and Morton, 1967) and the schemes developed by Gadd (1978, 1980) and Purnell (1976), more or less in increasing order of efficiency (De Haan, 1980).

The most usual finite difference schemes for the solution of the diffusion equation (cf. eq. IV.4), which has a parabolic form,

$$\partial c/\partial t = K\, \partial^2 c/\partial x^2 \tag{VIII.10}$$

are generally given by:

$$(c_j^{n+1} - c_j^n)/\Delta t = (K/\Delta x^2)[\theta(\delta^2 c)_j^{n+1} + (1-\theta)(\delta^2 c)_j^n], \tag{VIII.11}$$

where the operator, δ, is defined as $\delta c = c_{j+\frac{1}{2}} - c_{j-\frac{1}{2}}$.
Eq. (VIII.11) includes the (Euler) explicit scheme ($\theta=0$), and several implicit systems, for example $\theta=\frac{1}{2}$ (Cranck-Nicolson), $\theta=1$ and $\theta=\frac{1}{2}-\Delta z^2/(12K\Delta t)$.

In conclusion, it is remarked that the advent of larger computers makes the direct and accurate higher-order finite difference schemes more attractive than they used to be. The question, however, which method gives the best balance between accuracy and efficiency cannot be answered in general. The answer is strongly dependent on the type of model and the use of the model intended.

IX. REVIEW OF INTERREGIONAL TRANSPORT MODELS

In this section, a review is given of current transport models which cover transport distances from 50 - 5000 km. Each model is summarized separately. Owing to space limitations, not all details are included. In table IX.1, a summary is given of the listed models; a distinction is made with regard to the following aspects:

- pollutant
- scale
- type
- mathematical framework
- averaging time
- costs of operation

With exception of the costs of operation, these aspects have been discussed in the above chapters. The costs of operation are mainly determined by the input preparation efforts that are necessary, and the computer time and memory requirements. An estimate of these is hard to give; an impression of these requirements is obtained by considering the complexity of the input data and the mathematical framework of the model.

The complexity of the input data results from, e.g., the wind-field description and the inclusion of chemical reactions. The complexity of the mathematical framework increases in accordance whether analytical, Lagrangian, hybrid or Eulerian methods are applied. With these considerations in mind we have indicated the cost of the model in the last column of Table IX.1.

Table IX.1 Summary of models according to the classification given in section III. The symbols are explained in footnotes.

Model	Pollutant[1)]	Scale[2)]	Type[3)]	mathematical framework[4)]	averaging time[5)]	costs of operation[6)]
ADOM	3	s	c/d	e	e	m
AES	1	s	c/d	l	l	l
AIRSOX	2	s	c/d	h	l	l
ARL	1	s/m	c/d	h	e	l
ASTRAP	2	s	c/d	h	l	l
CEGB (SS)	2	m	c	e	e	l
CEGB (TD)	2	s	c/d	l	e	l
CIT	3	m	c	e	e	h
EPA	2	s	c/d	l	e	l
EPRI/ERT	3	s	c/d	e	e	h
EURMAP/ENAMAP	2	s	c/d	l	l	m
Fisher	2	s	d	a	l	l
Gillani	2	m	c/d	l	e	l
Hov	3	s	c/d	l	e	l
KNMI	3	m	c/d	e	e	h
Liraq	3	m	c	e	e	h
Mesoplume	2	s/m	c/d	l	e	l
Mesopuff	2	s/m	c/d	l	e	l
Mesogridd	2	s	c/d	e	e	m
Mesos	4	s/m	dos	l	e/l	l
OECD (ECE)	2	s	c/d	l	e/l	m
OME	1	s	c/d	l	l	l
Pseudo Spectral	2	s	c/d	e	e/l	l
RAPT	2	s	c/d	l	l	l
RCOM	2	s	c/d	a	l	l
RIV	1	m	c/d	l/e	e	l/m
SAI	3	m	c	e	e	h
SAI/EPA	2	s	c/d	e	e	m
STEM	3	s	c	e	e	h
Tald	4	s	dos	l	l	l
TDMB	1	m	c	e	e	m
Teknetron	2	s	c/d	e	e	l
TNO	2	m	c	a	l	l
Veltischeva	2	s	c/d	e	e	m
2 BPUFF	4	s/m	c	l	e	m

1) 1 = inert pollutants, 2 = chemically reactive pollutants modelled with conversion rates or decay constants, 3 = various pollutants including (photo)chemical reactions, 4 = radioactive pollutants

2) s = synoptic, m = mesoscale

3) c = determines ambient concentrations, d = deposition oriented, dos = dosage oriented

4) a = analytic, e = eulerian, h = hybrid, l = lagrangian

5) e = episodic, l = long term

6) l = low, m = moderate, h = high

ADOM model, Environmental Research and Technology, Concord, Mass, USA

References

Misra and Christie (1983).

Model type

Eulerian model describing transport and hydrocarbon/NO_x and SO_x chemistry.

Hor. scale	: 3500 - 7000 km (continental)
Time scale	: episodic
Hor. resolution	: 33 - 65 km
Vert. resolution	: presently 1 layer with extension to 12 and 20 layer versions.
Num. method	: hor. advection from interpolated Eulerian splines, flux corrected vert. diffusion from fully implicit scheme.

Windfield

Derived from a planetary boundary layer model

Turbulent Diffusion

Vertical : boundary layer scaling (Brost and Wyngaard, 1978).
Horizontal: ~ velocity deformation tensor according to Smagorinsky.

Deposition and Chemistry

Dry deposition: resistance type model based on stability and canopy resistance.
Wet deposition: rain and washout effects to be compared from cloud module and heterogeneous chemistry. Removal rates will be functions of space, time etc.
Chemistry : interchangeable chemical kinetic mechanism describing SO_2, NO_x and sulphate and nitrate aerosol formation.

Input Requirements

Emission inventories of NO_2 and HC's. Concentration distributions of some gas and aerosol reacting components. Wind, humidity, cloud, precipitation and validation data with a horizontal resolution of ~ 150 km for 6-12 vertical layers.

AES model, Atmospheric Environment Service, Downsview, Ontario, Canada

References

Olson et al. (1979), Voldner et al. (1979).

Model Type

A Lagrangian trajectory model which describes the transport and deposition of SO_2 and sulphate.

Hor. scale : ~ 2000 km (N.E.-U.S.A. + Canada)
Time scale : monthly to annual
Hor. resolution : ~ 300 km
Vert. resolution: one uniformly mixed layer with thickness dependent on monthly climatology.

Windfield

Objective analysis at 4 levels on 381 x 381 km grid, at 6 hourly intervals.

Turbulent Diffusion

Vertical : instantaneous mixing.
Horizontal : ?

Deposition and Chemistry

Dry deposition : deposition velocity for SO_2 and $SO_4^=$, 0.5 and 0.1 cm/s respectively.
Wet deposition : proportional to 24 hr precipitation amount.
Chemistry : SO_2-SO_4 conversion is 1% hr^{-1}.

Input Requirements

Meteorology : radiosonde and precipitation data.
Air pollution : emission inventory.

AIRSOX model, Brookhaven National Laboratories, Upton, New York 11973, USA

References

Meyer et al. (1979).

Model Type

Trajectory model for the description of dispersion and conversion of SO_2 and sulphate.

Hor. scale : ~ 4000 km (U.S.A.)
Time scale : month
Hor. resolution : ~ 30 km
Vert. resolution: 12 layers
Num. method : mod. Cranck-Nicholson finite difference scheme for vertical diffusion

Windfield

The horizontal transport is described by a wind velocity averaged in the vertical through the mixing layer.

Turbulent Diffusion

Vertical : eddy diffusivity concept.
Horizontal : gaussian profile.
Mixing height : 1000 m.

Deposition and Chemistry

Dry deposition : $V_d(SO_2) = 3\ 10^{-2}\ ms^{-1}$, V_d(sulphate) = $3\ 10^{-3}\ ms^{-1}$.
Wet deposition : included, not specified.
Chemistry : conversion rate for SO_2 sulphate $5\ 10^{-3}\ hr^{-1}$.

Input Requirements

Meteorology : upper air wind observations.
Air pollution : gridded emissions for SO_2 and sulphate.

ARL model, AIR Resources Laboratories, Silver Spring, Maryland, U.S.A.

References

Heffter et al. (1975); Draxler (1979); Heffter (1980).

Model Type

Two related trajectory models for inert pollutants: ATAD for synoptical scale and a mesoscale version (MS).

Hor. scale : ATAD: ~ 2000 km, MS: ~ 100 km
Time scale : episodic
Hor. resolution : -
Vert. resolution: ATAD: one transport layer
MS: 20 layers
Num. method : MS: finite difference method

Windfield

ATAD : trajectory is based on $1/r^2$ interpolation of upper air wind data.
MS : trajectory is based on $1/r^2$ interpolation of surface wind data, corrected according to observed upper air wind profiles.

Turbulent Diffusion

Vertical : ATAD: instantaneous mixing in transport layer; transport layer is equal to the highest mixing layer that occurred during previous transport. MS: eddy diffusivity K_z is derived from Pasquill Stability Classification, $K_z(P)$. When $z < 150$ m, $K_z = K_z(P)*z/150$. Above 150 m, $K_z = K_z(P)$. When $z > H$, K_z is reduced to 10^{-2} m^2/s.
Horizontal : gaussian ($\sigma_y \alpha t$).
Mixing height : variable, determined from radiosonde data during day time. At night H is taken equal to $2\sigma_z$, or $\alpha t^{\frac{1}{2}}$. A minimum value for H is 150 m.

Deposition and Chemistry

Dry deposition : ATAD: $V_d = 10^{-2} ms^{-1}$.
MS: not specified.

Wet deposition : ATAD: removal rate proportional to precipitation rate
MS: not specified.
Chemistry : not included.

Input Requirements

Meteorology : ATAD: upper air wind and temperature data.
MS: surface and upper air wind data, Pasquill stability data.
Air pollution : emissions.

ASTRAP, Argonne National Laboratory, Argonne, Illinois 60439, U.S.A.

References

Shannon, J.D. (1981): A model of regional long-term average sulphur atmospheric pollution, surface removal, and net horizontal flux. Atmospheric Environment (15 (5), 689-701).

Model Type

Statistical trajectory model for calculation of SO_2/SO_4 concentration and deposition, regional sulphur budget, NO_x/nitrate concentrations and deposition (preliminary).

Hor. scale : 50 - 2500 km
Time scale : long term (monthly to annual)
Hor. resolution : dependent upon resolution of meteorological and emission data-typically 50-100 km
Vert. resolution: 9 layers, thinnest 100 m, total depth 1800 m
Num. method : Gaussian moment-conservation for vertical diffusion.

Windfield

Time series of objective analyses produced by r^{-2} interpolation of average wind through 1.8 km depth at rawinsonde stations.

Turbulent Diffusion

Vertical : eddy diffusivity; diurnal pattern of K_z profile specified for each season.
Horizontal : Gaussian function fitted to distribution of trajectory endpoints as a function of plume age for a grid of virtual sources.
Mixing height : Implicit from K_z profiles; typical cycle of nocturnal inversion formation, intensification, lifting and erosion from below, and dissipation.

Deposition and Chemistry

Dry deposition : diurnally and seasonally varying pattern of deposition velocity; similar magnitudes for SO_2 and SO_4; diurnal averages ca. 0.4 cm s^{-1} (summer), 0.25 cm s^{-1} (winter).

Wet deposition : bulk removal of sulphur = $(h/10)^{\frac{1}{2}}$, where h is the 6-hrly precipitation in mm; maximum removal any 6-hr period 80%; rainfall equivalent used when precipitation is snow.

Chemistry : diurnally and seasonally varying linear first-order $SO_2 \rightarrow SO_4$ transformation; summer average 1.1% hr^{-1} (0.2 to 3.0% hr); winter average 0.48% hr^{-1} (0.1 to 1.5% hr^{-1}); augmented initial transformation in urban areas).

Input Requirements

Meteorology : Time series of rawinsonde winds.
Time series of 6-hrly precipitation observations.

Emissions : SO_x emissions by location, fuel and source type, and stack parameters.

CEGB model, Central Electricity Generating Board, Ratcliffe on Soar, Nottingham, U.K.

References

Maul (1979), Maul (1980).

Model Type

Two related models: an analytical steady-state model (SS) for SO_2 concentration calculations and a time dependent puff model (TD) for calculation of SO_2 and sulphate concentration.

Hor. scale	: SS: ~ 100 km; TD: ~ 1000 km.
Time scale	: episodic.
Hor. resolution	: SS: ~ 10 km; TD: ~ 100 km.
Vert. resolution	: analytical expressions for the vertical concentration profile.

Windfield

SS: uniform.
TD: based on surface pressure measurements.

Turbulent Diffusion

Vertical	: analytical solutions of the diffusion equation based on Green's functions; in TD calculated in time steps.
Horizontal	: SS: gaussian dispersion; TD: constant eddy coefficient, $K_H = 10^4\ m^2\ s^{-1}$.
Mixing height	: SS: constant, according to stability class. TD: variable, according to wind speed, radiation, surface roughness and other factors.

Deposition and Chemistry

Dry deposition	: SS: not included. TD: scavenging coefficient proportional to precipitation rate.
Chemistry	: SS and TD: constant conversion rate of $SO_2 \rightarrow$ sulphate, initial conversion of -10%.

Input Requirements

Meteorology : SS: wind data, stability class.
TD: surface pressure data, precipitation data, radiation, surface roughness.

Air pollution : SO_2 source emissions.

CIT model, California Institute of Technology, Pa., California 91125, U.S.A.

References

McRae et al. (1979).

Model Type

Eulerian grid model for calculation of dispersion and chemical transformation of a relatively large set of species.

Hor. scale : ~300 km (California Basin)
Time scale : episodic
Hor. resolution : a few km
Vert. resolution: several layers
Num. method : fourth order finite difference combined with non-linear filtering (Long and Hicks, 1975) for advection, centered difference for diffusion; Lear's method (Linger and Willoughby, 1967) for chemistry.

Windfield

Observations of surface and upper air winds are interpolated according to $1/r^2$. The wind field obtained is smoothed and the divergence of the horizontal flow field is reduced to an acceptable level.

Turbulent diffusion

Vertical : surface layer expressions based on u_* and L obtained from the scheme of Golder (1972). In the mixed layer, expressions from Lamb and Durran (1977), Shir (1973) and O'Brien (1970) for unstable, neutral and stable conditions respectively. Dispersion of point source emissions is treated gaussian close to the source.
Horizontal : gaussian close to the source, otherwise not specified.
Mixing height : variable according to input specifications.

Deposition and Chemistry

Dry deposition : deposition velocity dependent on surface layer parameters.

Wet deposition : -
Chemistry : 50 step, lumped hydrocarbon kinetic mechanism.

Input Requirements

Meteorology : surface and upper air wind and temperature data, mixing heights, stability class, surface roughness, UV intensity.
Air pollution : emissions per component of point and area sources, initial and boundary conditions per component.

EPA model, Environmental Sciences Laboratory, Research Triangle Park, N.C. 27711, U.S.A.

References

Henmi and Reiter (1979); Henmi (1980).

Model Type

Two layer segmented gaussian plume trajectory model for SO_2 and sulphate.

Hor. scale	: ~ 1500 km (Ohio River Basin)
Time scale	: 24-hour average
Hor. resolution	: ~ 100 km
Vert. resolution	: daytime 1 layer, nighttime 2 layers

Wind field

Layer averaged horizontal wind fields are derived from externally calculated wind field for (i) the nighttime layer h_1, (ii) the daytime mixed layer h_2 and (iii) the complementary layer, $h_2 - h_1$.

Turbulent Diffusion

Vertical	: instantaneous mixing within layer; no exchange between (nighttime) layers.
Horizontal	: dispersion coefficients derived from the vertical wind shear within each layer.
Mixing height	: h_1 (daytime): 1600 m; h_2 (nighttime): 240 m.

Deposition and Chemistry

Dry deposition	: $V_d(SO_2) = 2\ 10^{-2}\ ms^{-1}$, V_d (sulphate) = $4\ 10^{-3}\ ms^{-1}$.
Wet deposition	: proportional to precipitation rate.
Chemistry	: conversion rate of $SO_2 \rightarrow$ sulphate $10^{-1} - 10^{-2}\ hr^{-1}$.

Input Requirements

Meteorology	: radiosonde wind data, precipitation data.
Air pollution	: SO_2 emissions of 60 largest sources.

EPRI/ERT model, Environmental Research and Technology Inc., Santa Barbara, California 93103, U.S.A.

References

Lavery et al. (1979).

Model Type

Eulerian grid model for calculation of SO_2 and sulphate concentrations.

Hor. scale	: ~ 2000 km (Eastern U.S.A.)
Time scale	: 24-hour averages
Hor. resolution	: 80 km
Vert. resolution	: 3 layers (0/300/700/1500 m)
Num. method	: method of moments (Egan and Mahony, 1972)

Windfield

Objective analysis of radiosonde data. It is made non-divergent through a procedure adopted from Endlich (1967). Two temporal interpolation options.

Turbulent diffusion

Vertical	: eddy diffusivity surface layer expressions based on L and u_* derived from wind speed and radiosonde temperature data. These are extrapolated to the mixing height according to O'Brien (1970).
Horizontal	: instantaneous mixing within grid cell.
Mixing height	: from observations or some current expressions (Zilitinkevich, 1972; Wyngaard, 1975).

Deposition and chemistry

Dry deposition	: stability and roughness dependent deposition velocity (Weseley and Hicks, 1977).
Wet deposition	: -
Chemistry	: conversion to sulphate from kinetic mechanisms of 20 reactions. For the time being a conversion rate of $7\ 10^{-3}\ hr^{-1}$ is used.

Input Requirements

Meteorology	: surface and upper air wind and temperature data, surface roughness.
Air pollution	: seasonally averaged point and area source emissions, diurnal variations, hourly power plant emissions for nine components, initial and boundary conditions.

EURMAP and ENAMAP, Stanford Research Institute, Menlo Park, California, U.S.A.

References

Johnson et al. (1978); Bhumralkar et al. (1979); Bhumralkar et al. (1980).

Model Type

Three Langragian models for SO_2 and sulphate: EURMAP I, EURMAP II and ENAMAP.

Hor. scale : ~2500 km; W-Europe (EURMAP I, EURMAP II) and EN-U.S.A. (ENAMAP I).
Time scale : long term (EURMAP I, ENAMAP I) and short term (EURMAP II).
Hor. resolution : ~ 70 km.
Vert. resolution: one cell up to mixing height or puff height.

Windfield

EURMAP I : interpolated 850-mbar observations are reduced in speed by 25% and backed by 15°.
EURMAP II : based on surface and 850-mbar observations and a power law.
ENAMAP I : derived from 850-mbar observations.

Turbulent Diffusion

Vertical : EURMAP I and ENAMAP I: instantaneous mixing up to mixing height.
EURMAP II: instantaneous mixing between surface and 400 m height, with subsequent growth of upper boundary up to at most 1500 m.
Horizontal : EURMAP I and ENAMAP I: Fickian diffusion, K_H = 36 km^2 hr^{-1}.
EURMAP II: based on horizontal deformation of the wind field.
Mixing height : EURMAP I : 1000 m.
EURMAP II: diurnal variation, based on upper air surroundings.
ENAMAP I : seasonal variation between 1150 and 1500 m.

Deposition and Chemistry

Dry deposition	: EURMAP I : SO_2 0.029 hr^{-1}, sulphate 0.007 hr^{-1}.
	EURMAP II: SO_2 0.029/z_p hr^{-1}, sulphate 0.007/z_p hr^{-1}, with z_p puff height in km.
	ENAMAP I : SO_2 0.037 hr^{-1}, sulphate 0.007 hr^{-1}.
Wet deposition	: EURMAP I, II: SO_2 0.216 R hr^{-1}, sulphate 0.07 R hr^{-1}, with R precipitation rate in mm/hr.
	ENAMAP I: SO_2 0.28 R hr^{-1}, sulphate 0.07 R hr^{-1}.
Chemistry	: conversion rate $SO_2 \rightarrow$ sulphate 0.01 hr^{-1}.

Input Requirements

Meteorology	: 850-mbar wind observations, precipitation data, further for EURMAP II surface wind observations, upper air temperature data, surface level atmospheric stability data.
Air pollution	: EURMAP I : gridded annual SO_2 emissions.
	EURMAP II: gridded SO_2 emissions, including a diurnal variation.
	ENAMAP I : gridded SO_2 emissions, including a diurnal and seasonal variation.

Fisher model, Central Electricity Generating Laboratories, Leatherhead, Surrey KT22 7SE, Great Brittain

References

Fisher (1975); Fisher (1978).

Model Type

Long term deposition model for SO_2 and sulphate containing analytical expressions based on statistics of wind, stability and precipitation.

Hor. scale : ~ 2500 km (Europe)
Time scale : long term
Hor. resolution : 127 km
Vert. resolution: analytical expression for the vertical concentration profile

Windfield

Homogeneous and stationary during transport, based on surface observations.

Turbulent Diffusion

Vertical : analytical expression for the vertical concentration profile, based on dep. velocity and a height independent eddy diffusivity coefficient, which depends on meteorological class.
Horizontal : implicit.
Mixing height : variable according to wind speed and stability class, constant during transport.

Deposition and Chemistry

Dry deposition : dep. velocity: SO_2 0.5 cm/s, sulphate 0 cm/s (reference height at top of surface layer).

Wet deposition : method based on Rodhe and Grandell (1972). Scavenging coefficient SO_2 10^{-4} s^{-1}, sulphate 10^{-4} s^{-1} (during precipitation). Correction for regional variation in precipitation.
Chemistry : initial conversion $SO_2 \rightarrow$ sulphate 10%; conversion rate 10^{-6} s^{-1}.

Input Requirements

Meteorology	: long term statistics of wind and atmospheric stability observations, precipitation statistics.
Air pollution	: gridded annual sulphur emissions.

Gillani model, Washington University, St. Louis, MO 63130, U.S.A.

References

Gillani (1978), Gillani and Husar (1976).

Model type

Quasi-steady Lagrangian model for SO_2 point source, which calcultates cross wind integrated concentrations. The output of 1-hour simulation of a steady state sub-model is input for the next 1-hour simulation.

Hor. scale : 250 km
Time scale : short term (daytime)
Hor. resolution : the model calculates integrated or Gaussian shaped cross wind distribution
Vert. resolution: variable (21 layers)
Num. method : "method of lines" for vertical diffusion.

Windfield

Air trajectories are model input.

Turbulent Diffusion

Vertical : eddy diffusivity coefficient: different formulations for different parts of mixing layer and for different stabilities.
Horizontal : eddy diffusivity coefficient.
Mixing height : according to input specification.

Deposition and Chemistry

Dry deposition : dep. velocity; proportional to solar radiation.
Wet deposition : not included.
Chemistry : conversion rate $SO_2 \rightarrow$ sulphate; proportional to solar radiation.

Input Requirements

Meteorology : air trajectories, mixing heights, solar radiation, surface wind, surface temperature, surface roughness.
Air pollution : point source emissions; vertical area source emissions.

KNMI model, Royal Netherlands Meteorological Institute, De Bilt

References

Van Dop et al. (1980); Van Dop and De Haan (1983).

Model Type

Grid model describing transport and diffusion with the surface inversion or mixed layer in Eulerian frame work. Lagrangian description of advection (without diffusion) above these layers.

Hor. scale : 500 km (Netherlands and surrounding countries)
Time scale : episodic
Hor. resolution : 20 km
Vert. resolution: 50 m
Num. method : pseudo spectral for advection, Cranck-Nicholson finite difference for diffusion

Windfield

Constructed from four sets of data: the 10 m synoptic wind, the surface geostrophic wind, the radio sonde 850 mbar wind, and tower and mast data. The vertical windprofile is a combination of the surface layer wind profile and the (assumed linear) geostrophic wind profile, using height dependent weighting factors. Tower data are included by $1/r^2$ interpolation.

Turbulent Diffusion

Vertical : in surface inversion or mixed layer: eddy diffusivity coefficient, based on Obukhov length and friction velocity derived from surface data (Brost and Wyngaard, 1978). Above these layers: no diffusion.
Horizontal : instantaneous mixing with grid cell.
Mixing height : variable in time and space, derived from Carson-Tennekes inversion rise model (unstable) or according to Nieuwstadt (1981) (stable).

Deposition and Chemistry

Dry deposition : canopy dependent velocity, values are adopted from Sehmel (1980).
Wet deposition : derived from the precipitation pattern.

Chemistry : first order conversion rate $SO_2 \rightarrow$ sulphate, or (photo)chemical kinetic mechanism can be included.

Input Requirements

Meteorology : surface synoptic data of wind, temperature, pressure and precipitation; 850 mbar wind data are obtained from weather forecast analysis of radio sonde data.

Air pollution : detailed emission inventory with a resolution a 10 x 20 km^2.

Hov model, Norwegian Institute for Air Research, Lilleström, Norway

References

Eliassen et al. (1982).

Model Type

Trajectory model for NO_x, HC and SO_2.
Hor. scale : continental (Europe)
Time scale : 6 hourly
Hor. resolution : ~ 150 km
Vert. resolution : one layer up to mixing height (daily determined from 1200 GMT sondes).

Windfield

850 mb trajectories followed for 96 h each 6 hours, 6 determined from ~ 120 radiosonde stations within the area. Trajectories could alternatively be backed by 15° and windspeeds be reduced to 90%.

Turbulent Diffusion

Vertical : complete mixing up to boundary layer height.
Horizontal : none.
Mixing height : from 1200 GMT radiosonde data interpolated to 150 km grid.

Deposition and Chemistry

Dry deposition : deposition velocity concept.
Wet deposition : parameterized using relative humidity.
Chemistry : reaction mechanism of ~ 100 chemical reactions and 40 different species including ozone.

Input Requirements

Radiosonde data; emission data of SO_2, NO_x and HC's on 150 x 150 km^2 grid.

Liraq model, Lawrence Livermore National Laboratory, Livermore 94550, U.S.A.

References

MacCracken et al. (1978), Duewer et al. (1978) and Duewer et al. (1980)).

Model Type

Eulerian model, which describes the advection of vertically averaged concentration profiles.

Hor. scale	: 100-200 km
Time scale	: episodic (hourly)
Hor. resolution	: a few km
Vert. resolution	: analytical expression for the vertical concentration profile.
Num. method	: SHASTA (Boris and Book, 1973) for advection; first order upstream difference scheme (Molenkamp, 1968) when photochemistry is included.

Windfield

Obtained from surface data; for vertical extension a power law expression is used. A method to generate a non-divergent field is used (Dickerson, 1978).

Turbulent Diffusion

Vertical	: a concentration profile is prescribed and fitted to source strength and deposition conditions.
Horizontal	: described by the introduction of a horizontal diffusion velocity concept.
Mixing height	: variable in space and time, based on observations.

Deposition and Chemistry

Dry deposition	: species dependent deposition velocities.
Wet deposition	: -
Chemistry	: a kinetic reaction mechanism for 20 species can be used.

Input Requirements

Meteorology	: surface wind data, inversion base heights, solar radiation.
Air pollution	: gridded hourly surface and elevated emissions.

Mesoplume, Mesopuff, Mesogrid, Environmental Research and Technology, Concord, MA 01742, U.S.A.

References

Bass et al. (1979).

Model Type

Three models of increasing complexity for the description of dispersion and conversion of SO_2 and sulphate.
Mesoplume is a gaussian plume model, where the centre of the plume coincides with a calculated trajectory (segmented plume model).
Mesopuff is a trajectory model with gaussian puff dispersion (puff superposition model).
Mesogrid is an Eulerian model, which treats horizontal advection and vertical diffusion only.

Hor. scale	: 1000 km ("four corners region"); smaller scale possible for Mesoplume and Mesopuff.
Time scale	: episodic (24-hour average concentrations)
Hor. resolution	: 40 km
Vert. resolution	: Mesoplume, Mesopuff: gaussian dispersion formulation. Mesogrid: three layers.
Num. method	: Mesogrid: "methods of moments" for the advection, forward in time centered difference scheme for the vertical diffusion.

Windfield

Radio sonde data are used for the construction of horizontal wind fields at a desired level. Wind fields are obtained by linear interpolation in time and $1/r^2$ interpolation in space.
Mesoplume and Mesopuff use a "representative" uniform wind velocity.

Turbulent Diffusion

Vertical	: Mesoplume, Mesopuff: dispersion coefficients up to 100 km and thereafter taken proportional to $t^{\frac{1}{2}}$. Mesogrid: eddy diffusivity profile, K(z); for $z < 0.1$ H surface layer expressions are used; for $0.1 H < z < H$ an O'Brien expression is used. Monin-Obukhov stability is obtained from Golders conversion from Pasquill class.
Horizontal	: Mesoplume, Mesopuff: dispersion coefficients. Mesogrid: -

Mixing height : determined from radiosonde data (Holzworth's method), or taken proportional to the wind speed.

Deposition and Chemistry

Dry deposition : V_d (SO_2) = 10^{-2} ms^{-1}, V_d (sulphate) = 10^{-3} ms^{-1}.
Wet deposition : -
Chemistry : $SO_2 \rightarrow$ sulphate conversion rate of 2% hr^{-1}.

Input Requirements

Meteorology : radiosonde wind, temperature and pressure observations.

Air pollution : gridded SO_2 emissions and plume rise data, for Mesogrid initial and boundary conditions as well.

Mesos model, Imperial College of Science and Technology, London, U.K.

References

Apsimon et al. (1980); Apsimon and Goddard (1976); Apsimon et al. (1978).

Model Type

Trajectory model calculating dosages from a single source emission of a radioactive pollutant.

Hor. scale : ~ 2000 km (Europe)
Time scale : episodic or long term
Hor. resolution : 30-100 km
Vert. resolution : analytical profile within mixed layer; above mixed layer a series of stratified layers (100 m thick) is taken into account.

Windfield

The trajectory wind within the mixing height is based on the geostrophic wind field derived from surface pressure data and a power law.
Adjustments are made according to stability and roughness. The trajectory wind above the mixing height is geostrophic.

Turbulent Diffusion

Vertical : Initially gaussian dispersion based on Pasquill stability categories. Later the mixed layer is assumed well mixed. Above the mixed layer the vertical spread is neglected.
Horizontal : proportional to the travel time.
Mixing height : derived from the Carson-Smith-Tennekes model. Mechanical mixing is included. At night and over the sea some values based on stability are chosen.

Deposition and Chemistry

Dry deposition : deposition velocity with resistance term limiting deposition rate when vertical diffusivity is low.

Wet deposition : depends on intensity and nature of precipitation (deduced from "present weather" synop codes).

Chemistry : decay and conversion from one radionuclide into another can be incorporated.

Transformation : radioactive decay coefficient.

Input Requirements

Meteorology : surface data (pressure, temperature, humidity, precipitation, cloud cover, height of low cloud).

Air pollution : single source emission of up to 4 pollutants.

OECD (ECE) model, Norwegian Meteorological Institute, Oslo, Norway

References

Eliassen (1978); Eliassen en Saltbones (1975).

Model Type

Lagrangian model for SO_2 and sulphate.

Hor. scale	: ~ 3000 km (Western Europe)
Time scale	: long term (year) or short term (day)
Hor. resolution	: 127 km
Vert. resolution	: one cell up to mixing height

Windfield

Wind interpolated between 850-mbar wind observations, is directly used for trajectories.

Turbulent Diffusion

Vertical	: instantaneous mixing up to mixing height.
Horizontal	: instantaneous mixing in grid cell.
Mixing height	: 1000 m.

Deposition and Chemistry

Dry deposition	: initial deposition 15%; dep. velocity for conc. calculations: SO_2 0.65 cm s^{-1}, sulphate 0.4 cm s^{-1}; for deposition calculations: SO_2 0.8 cm s^{-1}, sulphate 0.2 cm s^{-1}.
Wet deposition	: for conc. calculations: SO_2 4 x 10^{-5} s^{-1}, sulphate 0 during precipitation; for deposition calculations: semi-empirical relation.
Chemistry	: conversion rate $SO_2 \rightarrow$ sulphate 3.5 x 10^{-6} s^{-1}.

Input Requirements

Meteorology	: 850-mbar wind observations, precipitation observations.
Air pollution	: annual total sulphur emissions on 127 x 127 km grid.

OME model, Ministry of the Environment, Toronto, Ontario, M5SIZ8, Canada

References

Venkatram et al. (1980).

Model Type

A lagrangian statistical model for SO_2 and sulphate.

Hor. scale	: ~ 2000 km (N.E. United States and Canada)
Time scale	: annual
Hor. resolution	: ~ 100 km
Vert. resolution	: one layer (1000 m thick).

Windfield

Long term wind statistics are applied.

Turbulent Diffusion

Vertical	: uniform distributions.
Horizontal	: proportional to travel time and statistics of 850 mb wind velocity.

Deposition and Chemistry

Dry deposition	: 0.5 cm/s for SO_2; 0.05 for $SO_4^=$.
Wet deposition	: dependent on average length of wet and dry periods.
Chemistry	: SO_2-$SO_4^=$ conversion rate dependent on dry (1% hr^{-1}) or wet conditions.

Input Requirements

Meteorology	: 850 mb winddata
Air pollution	: emission inventory (lumped).

Pseudo Spectral model, Danish Air Pollution Laboratory, Risø, 4000 Roskilde, Denmark

References

Prahm and Christensen (1977).

Model Type

Two dimensional Eulerian model describing the transport and conversion of SO_2 and sulphate.

Hor. scale : ~ 4000 km (Europe)
Time scale : 24 hour averages
Hor. resolution : 127 km
Vert. resolution : one cell up to mixing height
Num. method : advection by the pseudo spectral method, time integration by third order linear multistep method.

Windfield

850 mbar wind data are interpolated. Time interpolation is linear.

Turbulent Diffusion

Vertical : neglected.
Horizontal : constant eddy coefficient, $K_H = 10^4\ m^2 s^{-1}$.
Mixing height : 1000 m.

Deposition and Chemistry

Dry and wet deposition and $SO_2 \rightarrow$ sulphate conversion: modelled by a decay constant.

Input Requirements

Meteorology : 850-mbar wind field
Air pollution : gridded annual total sulphur emissions (OECD).

RAPT model, Pacific Northwest Laboratory, Richland, WA 99352, U.S.A.

References

McNaughton (1980), Powell et al. (1978), McNaughton et al. (1981 a,b).

Model Type

Trajectory model for SO_2, sulphate, particles, NO_2 and NO_3.

Hor. scale : variable; typically ~ 2000 km (NE-U.S.A.)
Time scale : long term (month)
Hor. resolution : ~ 170 km for transport calculation, ~ 34 km for dispersion, deposition and transformation calculations.
Vert. resolution : single transport layer (~ 100-1000 m).

Windfield

Objective interpolation of radiosonde data in the mixed layer.

Turbulent Diffusion

Optional; typically uniform within mixed layer.
Mixing height : sinusoidal diurnal cycle between 100 and 1000 m.

Deposition and chemistry

Dry deposition : dep. velocity, SO_2 1.0 cm s^{-1}, sulphate 0.1 cm s^{-1}

Wet deposition : SO_2: removal rate = 0.005 P(t) mm, with P(t) prec. rate (mm/hr^{-1}). sulphate: w = 0.232 $P(t)^{0.625}$ mm.

Chemistry : first order transformation rate $SO_2 \rightarrow$ sulphate, daytime 0.02 hr^{-1}, nighttime 0.0025 hr^{-1} (dependency on precipitation rate included).

Input Requirements

Meteorology	: upper air wind observations in 100 - 1000 m layer, hourly prec. data.
Air pollution	: time averaged emissions from major point sources.

Remark

The model has been used in assessment and is undergoing substantial verification testing.

RCDM-model, Teknetron Research Inc., Boston, Mass. 02154, U.S.A.

References

Fay and Rosenzweig (1980).

Model Type

Analytical Eulerian model using statistics of trajectories.
Hor. scale : continental
Vert. scale : seasonal value of mixing height.
Time scale : seasonally or yearly
Hor. resolution : ~ 50 km.

Windfield

Average vector wind field derived from upper air wind data.

Turbulent Diffusion

Horizontal : constant
Vertical : steady state analytical solution of two dimensional diffusion equation.

Deposition and Chemistry

Dry and wet deposition are combined for SO_2 and SO_4 respectively.
Chemistry : chemical conversion factor.

Input Requirements

Emission data; seasonal precipitation data.

RIV-model, National Institute of Public Health, Bilthoven, the Netherlands

References

Van Egmond and Kesseboom (1983a, 1983b), Klug et al. (1984).

Model Type

Both a Lagrangian puff model and an Eulerian grid model have been developed which describe the transport of SO_2.

Hor. scale : 400 km (the Netherlands)
Time scale : episodic
Hor. resolution : 10 km (grid model)
Vert. resolution : both models contain a surface-mixed- and reservoir layer. High sources emit in the reservoir layer.
Numerical method : advection is described separately in the mixed and reservoir layer. Within each layer perfect mixing is assumed (grid model).

Windfield

Derived from surface and tower wind observations. Data are fitted to an Ekman type wind profile.

Turbulent Diffusion

Vertical : Gaussian dispersion (puff model) in mixed layer.
Horizontal : Taylor's diffusion (puff model); constant horizontal diffusivity (grid model).
Mixing height : estimated from sodar and sonde data, if necessary Zilinkevich's expression is used for the stable boundary layer height.

Deposition and Chemistry

Dry deposition : based on stability and surface resistance.
Wet deposition : -
Chemical transformation: 1% hr^{-1} conversion factor.

Input Requirements

Emission inventory, local meteorological measurements.

SAI model, Systems Applications Inc., San Rafael, California 94903, U.S.A.

References

Reynolds (1979); Builtjes et al. (1980).

Model Type

3-dimensional grid model including photochemical reactions for up to 20 species, including SO_2, sulphate, NO and NO_2.

Hor. scale	: ~ 30 - 300 km
Time scale	: short term (day)
Hor. resolution	: ~ 2 - 10 km
Vert. resolution	: variable, ~ 5 layer
Numerical methods	: SHASTA method (Boris and Book, 1973) for advection, finite difference method for diffusion.

Windfield

Several options; direct interpolation of surface data and vertical profiles, or divergence free wind field including heat island effects.

Turbulent Diffusion

Vertical	: eddy diffusivity coefficient, according to Businger and Arya (1974) (stable conditions) and Deardorf and Willis (1975) (neutral and unstable conditions).
Horizontal	: eddy diffusivity 50 ms^{-1}.
Mixing height	: variable in time and space, according to input specifications.

Deposition and Chemistry

Dry deposition	: dep. velocity, variable per surface grid cell according to land use and per component.
Wet deposition	: not included.
Chemistry	: photochemical kinetic mechanism, based on the Carbon-Bond Mechanism; SO_2 conversion is only photochemical.

Input Requirements

Meteorology	: surface and upper air wind data, mixing heights, stability class, surface roughness, surface temperatures, vertical temperature gradients, humidity, atmospheric pressure, UV intensity.
Air pollution	: emissions per component of point and area sources, roadway emissions, initial and boundary conditions for 13 components.

SAI/EPA-model, Systems Applications, Inc. San Rafael, California 94903, U.S.A.

References

Liu and Durran (1977), Durran et al. (1979).

Model Type

3-dimensional grid model for SO_2 and sulphate; a surface layer model is embedded in a mixing layer model.

Hor. scale : ~ 1000 km (Northern Great Plains, U.S.A.).
Time scale : short term (days).
Hor. resolution : ?
Vert. resolution : 1 surface layer, 1 mixing layer.
Num. method : SHASTA method (Boris and Book, 1973) for advection.

Windfield

Above the surface layer: geostrophic 850 mbar wind.
In the surface layer : interpolated surface wind observations.

Turbulent Diffusion

Vertical : in mixing layer uniform vertical concentration profile. Surface layer is used for description of pollutant flux to the surface.
Horizontal : eddy diffusivity coefficient 10^4 m^2 s^{-1}.
Mixing height : variable according to input specifications.

Deposition and Chemistry

Dry deposition : dependent on surface characteristics and turbulence, based on formulation by Thom (1972).
Wet deposition : -
Chemistry : first order conversion rate $SO_2 \rightarrow$ sulphate, several values are used.

Input Requirements

Meteorology : 850-mbar pressure distribution, surface wind observations.
Air pollution : gridded SO_x emissions of point sources, initial and boundary conditions.

STEM-model, University of Iowa, Iowa (U.S.A.) 52242

References

Carmichael et al. (1979, 1984).

Model Type

Three dimensional Eulerian model including chemistry.

Hor. scale	: continental; ~ 100 km
Time scale	: hourly
Hor. resolution	: ~ 30 km
Vert. resolution	: 11 layers up to 8 km
Numerical methods	: locally one dimensional finite element methods; time split.
Windfield	: horizontal $1/r^2$ interpolation of surface wind-data. Vertical extrapolation by means of power law profile.

Turbulent Diffusion

Vertical	: based on Yamada and Mellor's (1975) b1 model.
Horizontal	: ?
Mixing height	: Yamada and Mellor (1975).

Deposition and Chemistry

Dry deposition	: dry deposition velocity concept.
Wet deposition	: ?
Chemistry	: extensive chemical kinetic mechanism, including photochemistry and cloud physics, and removal at the earth's surface.

Input Requirements

Wind velocity, temperature, humidity, cloud cover (specified), radiation, surface type, emissions.

Tald model, Institut de Protection et de Surêté Nucleaire, dept. de Protection, 92260 Fontenay-aux-Roses, France

References

Despres et al. (1980).

Model Type

Trajectory model for application to single source radioactive releases. The model calculates frequency distributions of dosages (time integrated concentrations).

Hor. scale : ~ 3000 km (W-Europe)
Time scale : long term
Hor. resolution : ~ 200 km
Vert. resolution : analytical concentration profile

Windfield

An average (925 mbar) wind velocity is used to calculate the trajectories. Spatial interpolation is done by a meteorological forecast model, that 00 and 12 GMT. Linear time interpolation is used. Close to the source, wind measurements in situ are used.

Turbulent Diffusion

Vertical : gaussian dispersion, uniform concentration profile, when σ_z equals the mixing height.
Horizontal : gaussian dispersion.
Mixing height : 1000 m.

Deposition and Chemistry

Dry deposition : deposition velocity depending on species.
Wet deposition : -
Chemistry : -
Transformation : radioactive decay coefficient.

Input Requirements

Meteorology : wind data obtained by the forecast model of the French Meteorological Office.
Air pollution : single source emission.

TDMB model, Technische Hochschule, Darmstadt, F.R.G.

References

Klug et al. (1978), Klug and Gerth (1980).

Model Type

Three dimensional Eulerian grid model which describes the dispersion of an inert pollutant (SO_2).

Hor. scale : ~ 200 km (F.R.G./The Netherlands), Münsterland
Time scale : episodic
Hor. resolution : 10 km
Vert. resolution : several layers
Num. method : Euler explicit difference scheme (diffusion and advection)

Windfield

Based on $1/r^2$ interpolation of surface and upper air wind data.

Turbulent Diffusion

Vertical : lapse rate dependent eddy diffusivity K_z.
Horizontal : instantaneous mixing within grid cell.
Mixing height : derived from a stability scheme.

Deposition and Chemistry

Dry deposition : deposition velocities vary depending on surface wind speed.
Wet deposition : included; modelled by a decay constant depending on precipitation intensity.
Chemistry : included; modelled by a decay constant.

Input Requirements

Meteorology : Surface and upper air wind and temperature data; furthermore special meteorological observations were done and used in the model.
Air pollution : gridded SO_2 emissions.

Teknetron model, Teknotron Inc., Boston, Massachussets 02154, U.S.A.

References

Niemann et al. (1979); Mills and Hirata (1979).

Model Type

Two-dimensional Eulerian model describing transport and conversion of SO_2/sulphate.

Hor. scale : ~ 1400 x 2000 km^2 (Eastern U.S.A.)
Time scale : episodic (daily averages)
Hor. resolution : 80 km
Vert. resolution : one layer up to the mixing height
Num. method : pseudo spectral advection scheme

Windfield

The transport is described by a representative wind velocity obtained by interpolation of radiosonde data for 600 m height. Linear time interpolation.

Turbulent Diffusion

Vertical : instantaneous mixing up to mixing height.
Horizontal : instantaneous mixing in cell.
Mixing height : variable, estimated from radiosonde data.

Deposition and Chemistry

Dry deposition : V_d (SO_2) = 1-2 10^{-2} ms^{-1},
V_d (sulphate) = 1-2 10^{-3} ms^{-1}.
Wet deposition : -
Chemistry : conversion rate for SO_2 → sulphate 1-2 10^{-2} hr^{-1}.

Input Requirements

Meteorology : radiosonde wind data, mixing heights.
Air pollution : SO_2 and sulphate emissions (SURE II inventory).

TNO model, Netherlands Organization for Applied Scientific Research TNO, Delft, the Netherlands

References

Van Egmond and Huygen (1979), Van den Hout (1981).

Model Type

Analytical plume model for SO_2 based on statistical distributions of wind speed, wind direction and dispersion category.

Hor. scale : ~ 500 km
Time scale : long term (season, year)
Hor. resolution : 1 - 25 km, according to source distribution
Vert. resolution : gaussian plume formulation for distance ≦ 50 km, constant concentration profile beyond 50 km.

Windfield

Uniform and stationary during transport, derived from surface observations, using a power law relation.

Turbulent Diffusion

Vertical : gaussian plume profile, based on Pasquill dispersion parameters; beyond 50 km uniform vertical profile; correction for steady state assumption.
Horizontal : implicit
Mixing height : depending on stability class; constant during transport.

Deposition and Chemistry

Dry deposition : dep. velocity 0.8 cm/s.
Wet deposition : removal rate 0.01 hr^{-1}.
Chemistry : conversion rate SO_2 → sulphate 0.01 hr^{-1}.

Input Requirements

Meteorology : annual (seasonal) statistics of surface wind and atmospheric stability class.
Air pollution : detailed emission inventory for point and area sources.

Veltischeva model, Hydrometeorological Research Centre of the U.S.S.R., Moscow

References

Veltischeva (1979).

Model Type

Eulerian grid model for SO_2 dispersion with inclusion of vertical wind.

Hor. scale	: 2000 km (Eastern Europe)
Time scale	: episodic
Hor. resolution	: 300 km
Vert. resolution	: 5-7 layers
Num. method	: Time split procedure, second order in time and space.

Windfield

Polynomial and linear interpolation of surface wind and 850 mbar data. Vertical wind is included.

Turbulent Diffusion

Vertical	: eddy diffusivity, primarily kept constant, but later varied with height.
Horizontal	: constant diffusion coefficient.
Mixing height	: 2000 m.

Deposition and Chemistry

Dry and wet deposition and conversion of SO_2: decay parameter.

Input Requirements

Meteorology	: surface and radiosonde wind data.
Air pollution	: gridded SO_2 emissions.

2BPUFF model, Lawrence Livermore National Laboratory, P.O. Box 808, Livermore, CA 94550

References

Crawford and Todd (1966), Knox et al. (1971) and Walton (1973).

Model Type

Two-dimensional, axially symmetric lagrangian model for calculating diffusion and deposition of particles and gases in a frame of reference moving with the center of a puff of effluent.

Hor. scale	: 10 to thousands of km
Time scale	: hours to days
Hor. resolutions	: < 1 km close in, tens of km at long range
Vert. resolutions	: hundreds of meters
Num. method	: finite difference solution of diffusion equation.

Windfield

From air trajectory at puff center; trajectory based on nearest constant pressure surface analyses and forecasts.

Turbulent Diffusion

Vertical	: K_z input for surface boundary layer, mixed layer, and free atmosphere above mixed layer. Input is time variable.
Horizontal	: Atmospheric dissipation rate (ε) at puff center is a time variable input parameter. Scale dependent horizontal diffusion (Walton, 1973) is used to determine K_h.
Mixing height	: Determined by time variable input values of the top of the mixed layer and the tropopause height.

Deposition and Chemistry

Dry deposition	: As particles reach lowest cell, they deposit according to input deposition velocity. Puff is depleted by an equivalent amount.

Wet deposition : Either by time variable input of precipitation rate or by vertical integral of puff activity. In both cases, the top of the precipitation systems is input.

Chemistry : Not included.

Input Requirements

Meteorology : Surface boundary layer height and wind speed, puff center wind speed, K_z in and above mixed layer, heights of mixed layer top and tropopause, precipitation rate, height of top of precipitation system, average wind speed along entire trajectory.

Air pollution : Source terms, half-lives, and dose conversion factors of radionuclides of concern.

The authors are indebted to many of the authors mentioned in section IX of this chapter for their helpful suggestions and corrections.

X. REFERENCES

Aalst, R.M. van and Bergsma, F. (1981). Removal and transformation processes in the atmosphere with respect to SO_2, NO_x NATO-CCMS, report No. 127; an updated version of this report is published in this Volume as Chapter 3.

Apsimon, H.M. and Goddard, A.J.H. (1976). Modelling the atmospheric dispersal of radioactive pollutants beyond the first few hours of travel. Procs. of the 7th ITM, NATO-CCMS, Airlie, Va., 7-10 September.

Apsimon, H.M., Earnshaw, D., Goddard, A.J.H. and Wrigley, J. (1978). Modelling of mesoscale atmospheric dispersion and the application of the mesos model to the 1957 wind scale release. Final Report Environmental Safety Group, SAF013.

Apsimon, H.M., Goddard, A.J.H. and Wrigley, J. (1980). Estimating the possible transfrontier consequences of accidental releases: The mesos model for long-range atmospheric dispersal. Procs. of a seminar on radioactive releases and their dispersion in the atmosphere following a hypothetical reactor accident, Risø, 22 - 25 April 1980 (CEC).

Bass, A., Benkley, C.W., Scire, J.S. and Morris, C.S. (1979). Development of mesoscale air quality simulation models, vol. 1 - 6, Environmental Research and Technology, Inc. Concord, MA 01742, EPA 600/7-79-XX, U.S.A.

Bhumralkar, C.M., Johnson, W.B., Mancuso, R.L. and Wolf, D.E. (1979). Region patterns and transfrontier exchanges of airborne sulphur pollution in Europe. Final report, SRI project 4797, SRI International, Menlo Park, California, U.S.A.

Bhumralkar, C.M., Mancuso, R.L., Wolf, D.E., Thullier, R.A., Nitz, K.C. and Johnson, W.B. (1980). Adaptation and application of a long-term air pollution model ENAMAP-I to Eastern North America. Final report, SRI project 7760, SRI International, Menlo Park, California, U.S.A.

Boris, J.P. and Book, D.L. (1973). Flux corrected transport. I. SHASTA, A fluid transport algorithm that works. Journ. Comp. Phys. 11, 38 - 69.

Brost, R.A. and Wyngaard, J.C. (1978). A model study of the stably stratified planetary boundary layer. J. Atmos. Sci., 35, 1427 - 1440.

Builtjes, P.J.H., Hout, K.D. van den, Veldt, C., Huldy, H.J., Hulshoff, J., Basting, W. and Aalst, R.M. van (1980). Application of a photochemical dispersion model to the Netherlands and its surroundings. Procs. of the 11th ITM, NATO-CCMS, Amsterdam. (Ed. C. de Wispelaere), Plenum Press, New York and London 1981, 621-37.

Businger, J.A. and Arya, S.P.S. (1974). Height of the mixed layer in the stably stratified planetary boundary layer, Advances in Geophysics, 18A, Academic Press, New York, 73 - 92.
Carmichael, G.R., Kitada, T. and Peters, L.K. (1984). A second generation combined transport/chemistry model for the regional transport of SO_x and NO_x compounds. Air Pollution Modelling and its Applications III (ed. C. de Wispelaere), p. 525-528, Plenum, New York.
Crawford, Todd V. "A Computer Program for Calculating the Atmospheric Dispersion of Large Clouds", Lawrence Livermore National Laboratory Report UCRL-50179, 1966.
Deardorf, J., and Willis, G.E. (1975). A parameterization of diffusion into the mixed layer. J. Appl. Meteor., 14, 1451 - 1458.
Despres, A., Doury, A. and Le Grand, J. (1980). Le programme de calcul Tald pour l'evaluation des transferts atmosphériques a longue distance dans les basses couches de l'atmosphere. Procs. of the seminar on radioactive releases and their dispersion in the atmosphere following a hypothetical reactor accident, Risø, 22 - 25 April 1980 (CEC).
Dickerson, M.H. (1978). MASCON - A mass consistent atmospheric flux model for regions with complex topography. J. Appl. Meteor., 17, 241 - 253.
Dop, H. van, Haan, B.J. de and Cats, G.J. (1980). Meteorological input for a three dimensional medium range air quality model. Procs. of the 11th ITM, NATO-CCMS, 24 - 27 November, Amsterdam, The Netherlands.
Dop, H. van and Haan, B.J. de (1983). Mesoscale Air Pollution Dispersion Modelling. Atmos. Environment 17, 1449-1456.
Draxler, R.R. (1979). Modelling the results of two recent mesoscale dispersion experiments, Atmospheric Environment 13, 1523 - 1533.
Duewer, W.H., MacCracken, M.C. and Walton, J.J. (1978). The Livermore Regional Air Quality Model: II Verification and Sample Application in the San Francisco Bay Area. J. Appl. Meteor. 17, 273-311.
Duewer, W.H., Walton, J.J., Grant, K.E. and Walker, H. (1980). Livermore Regional Air Quality Model (LIRAQ) Transfer to EPA. Report No. UCRL-52 864 (available from NTIS). Also published by EPA.
Durran, D.R., Meldgin, M.J. and Liu, M.K. (1979). A study of long range air pollution problems related to coal development in the Norhtern Great Plains. Atmospheric Environment 13, 1021-1087.
Egan, B.A. and Mahoney, J.R. (1972). Numerical modelling of advection and diffusion of urban area source pollutants, Journ. Appl. Meteor. 11, 312 - 322.
Egmond, N.D. and Huygen, C. (1979). Evaluation of a meso-scale model of sulphur dioxide. Procs. of the 10th NATO/CCMS Int. Technical Meeting on air pollution modelling and its applications, Rome, Italy.

Egmond, N.D. van and Kesseboom, H. (1983a). Mesoscale Air Pollution Dispersion Models - I Eulerian Grid Model. Atmos. Environment 17, 257-265.

Egmond, N.D. van and Kesseboom, H. (1983b). Mesoscale Air Pollution Dispersion Models - II Lagrangian Puff Model and Comparison with Eulerian Grid Model. Atmos. Environment 17, 267-274.

Eliassen, A. and Saltbones, J. (1973). A one layer Lagrangian model for the description of long-range transports of air pollutants. Expert meeting on long-range transport of air pollutants, Gansdal, 1973, Norwegian Institute for Air Research.

Eliassen, A. and Saltbones, J. (1975). Decay and transformation rates of SO_2, as estimated from emission data, trajectories and measured air concentrations. Atmospheric Environment 9, 425-429.

Eliassen, A. (1978). The OECD study on long range transport of air pollutants: Long range transport modelling. J. Appl. Meteor. 12, 479-487.

Eliassen, A. (1980). A review of long-range transport modelling, J. Appl. Met. 19, 231-240.

Eliassen, A., Hov, Ø., Isaksen, I.S.A., Saltbones, J. and Stordal, F. (1982). A Lagrangian long-range transport model with atmospheric boundary layer-chemistry. Journal of Appl. Meteor. 11, 1645-1661.

Endlich, R.M. (1967). An iterative method for altering the kinematic properties of wind field, Appl. Meteor., 6, 837-844.

Fay, J.A. and Rosenzweig, J.J. (1980). An Analytical Diffusion Model for Long Distance Transport of Air Pollutants. Atmos. Environment 14, 355-365.

Fisher, B.E.A. (1975). The long range transport of sulphur dioxide. Atmospheric Environment 9, 1063-1070.

Fisher, B.E.A. (1978). The calculation of long term sulphur deposition in Europe. Atmospheric Environment 12, 489-501.

Gadd, A.J. (1978). A numerical advection scheme with small phase speed errors. Qu. J. Roy. Met. Soc. 104, 583-594.

Gadd, A.J. (1980). Two refinements of the split explicit integration scheme. Qu. J. Roy. Met. Soc. 108, 215-220.

Gillani, N.V. (1978). Project MISST: Mesoscale plume modeling of the dispersion, transformation and ground removal of SO_2. Atmospheric Environment 12, 569-588.

Gillani, N.V. and Husar, R.B. (1976). Analytical-Numerical Model of Air Pollutants. Proceedings of the 7th NATO/CCMS Int. Tech. Meeting on Air Pollution Modeling and Its Application, Airlie, Virginia, Sept. 1976.

Golder, D. (1972). Relations among stability parameters in the surface layer. Bound. Layer Meteor., 3, 47.

Granat, L., Rodhe, H. and Hallberg, R.O. (1976). The global sulphur cycle. In: Svensson, B.H. and R. Söderlund (eds.) Nitrogen, Phosphorus and Sulphur - Global Cycles. SCOPE Report 7. Ecol. Bull. (Stockholm) 22, 89-134.

Haan, B.J. de (1980). A comparison of finite difference schemes describing the two-dimensional advection equation. Procs. of the 11th ITM, NATO/CCMS, Amsterdam 24 - 27 November.

Heffter, J.L., Ferber, G.J. and Taylor, A.D. (1975). A regional-continental scale transport, diffusion and deposition model. NOAA technical memorandum ERL-ARL-50, 28 p.

Heffter, J.L. (1980). Air resources laboratories atmospheric transport and dispersion model (ARL-ATAD), NOAA technical memorandum ERL-ARL-81, 17 p.

Henmi, T. (1980). Long-range transport model of SO_2 and sulphate and its application to the Eastern United States. J. of Geophys. Research 85, 4436-4442.

Henmi, T. and Reiter, E.R. (1979). Long-range transport and transformation of SO_2 and sulphate, E.P.A. report 600/4-79-068, 168 p.

Holzworth, G.C. (1972). Mixing heights, wind speeds and potential for urban pollution throughout the contiguous United States, Publ. AP-101, U.S. Environ. Prot. Agency, Research Triangle Park, N.C.

Hout, K.D. van den (1981). Model calculations of ambient SO_2 and NO_x concentration patterns in the Netherlands. Procs. IVth Int. Symp. on Industrial Chimneys, The Hague, Netherlands.

Johnson, W.B., Wolf, D.E. and Mancuso, R.L. (1978). Long term regional patterns and transfrontier exchanges of airborne sulphur pollution in Europe. Atm. Env. 12, 511-527.

Klug, W. and Gerth, (1980) 3. Arbeitsbericht RIM-project, Umweltbundesamt Berlin, 110 p.

Klug, W., Dietzer, B. and Hermann, K. (1978). Interregional Transport von Luftverunreinigungen, Arbeitsbericht 12, Umweltbundesamt, Berlin, 74 p.

Knox, J.B., Crawford, T.V., Peterson, K.R. and Crandall, W.K. "Comparison of U.S. and USSR Methods of Calculating the Transport, Diffusion, and Deposition of Radioactivity", Lawrence Livermore National Laboratory Report UCRL-51054, 1971.

Lange, R. (1978). ADPIC - A three-dimensional particle-in cell model for the dispersal of atmospheric pollutants and its comparison to regional tracer studies, Journ. Appl. Meteor. 17, 320-329.

Lavery, T.F., Thrasher, J.W., Godden, D.A., Lloyd, A.C. and Hidy, G.M. (1979). Regional transport and photochemical model of atmospheric sulphates. Procs. of the 9th ITM, NATO/CCMS, Toronto, 28-31 August, p. 353-362.

Linger, W. and R.A. Willonghby (1967). Efficient numerical integration of stiff systems of ordinary differential equations. IBM research report no. RL-1970.

Liu, M.K. and Durran, D.R. (1977). The development of a regional air pollution model and its application to the Northern Great Plains. E.P.A.-908/1-77-001, Office of Energy Activities, US-EPA, Denver, Colorado.

Long, P.E. and Hicks, F.J. (1975). Simple properties of Chapeau function and their application to the solution of the advection equation. TDLR Report no. TDL 75-8. Silver Springs, MD.

Maul, P.R. (1979). The mathematical modelling of meso-scale and long-range transport of sulphur compounds. Procs. of the 10th NATO/CCMS International Technical Meeting on air pollution modelling and its applications, Rome, Italy, October 22 - 26.

Maul, P.R. (1980). Atmospheric transport of Sulphur Compound Pollutants. Ph. D. thesis. Imperial College, London. C.E.G.B. internal report MID/SSD/80/0026/R.

McCracken, M.C. et al. (1978). The Livermore regional air quality model: I. Concept and development, Journ. Appl. Meteor. 17, 254-272.

MacNaughton, D.J. (1980). Initial comparison of SURE/MAP3S sulphur oxide observations with long-term regional model predictions. Atmospheric Environment 14, 55 - 63.

MacNaughton, D.J. and Powell, D.C. (1981a). RAPT - The Pacific Northwest Laboratory Regional Air Pollutant Transport Model: A Guide. U.S. EPA, Battelle Pacific Northwest Lab., PNL-3390, Richland, WA 99352.

MacNaughton, D.J., Berkowitz, C.M. and Williams, R.C. (1981b). A diagnostic Analysis of a Long Term Regional Air Pollutant Transport Model. J. Appl. Meteor. 20, 795-201.

McRae, G.J., Goodin, W.R. and Seinfeld, J.H. (1979). Development of a second-generation airshed model for photochemical air pollution. Procs. of the 4th symposium on turbulence, diffusion and air pollution, Reno, 15 - 19 January, AMS 389 - 396.

Mesinger, F. and Arakawa, A. (1976). Numerical methods used in atmospheric models. Garp publication series no. 17.1 World Meteorological Organization (WMO), Geneva.

Meyers, R.E., Cederwall, R.T., Storch, J.A. and Kleinman, L.I. (1979). Modelling sulphur oxide concentrations in the Eastern United States: model sensitivity, verification and applications. Procs. of the 4th symposium on turbulence, diffusion and air pollution, Reno, 15 - 16 January, p. 673-676.

Mills, M.T. and Hirata, A.A. (1979). A multiscale dispersion model for local and regional $SO_2/SO_4^=$ concentrations: formulation and initial evaluation. Procs. of the 9th ITM, NATO/CCMS, 23-26 October, Rome, Italy, p. 397-406.

Misra, P.K. and Christie, A.D. (1983). Acid Deposition and Oxidant Model. Procs. of the EPA-OECD International Conference on Long Range Transport Model for Photochemical Oxidants and their Precursors; Report no. 600/9-84.006 Research Triangle Park, NC.

Molenkamp, C.R. (1968). Accuracy of finite difference methods applied to the advection equation. J. Appl. Meteor., 7, 160-167.

Niemann, B.L., Hirata, A.A. and Smith, L.E. (1979). Application of a regional transport model to the simulation of multi-scale sulphate episodes over the Eastern United States and Canada. Procs. of the WMO symposium on the long-range transport of pollutants. Sofia, 1-5 October. WMO report no. 538.

Nieuwstadt, F.T.M. (1981). The steady-state height and resistance laws of the nocturnal boundary layer: theory compared with Cabauw observations. Boundary-Layer Meteorol., 20, 3-17.

Nieuwstadt, F.T.M. and Haan, B.J. de (1981). An analytical solution of the one-dimensional diffusion equation in a non-stationary boundary layer with an application to inversion rise fumigation. Atm. Envir. 15, 845-851.

O'Brien, J.J. (1970). A note on the vertical structure of the eddy exchange coefficient in the planetary boundary layer. J. Atmos. Sci., 17, 1213-1215.

Olson, M.P., Voldner, E.C., Oikawa, K.K. and Macafee, A.W. (1979). A concentration/deposition model applied to the Canadian long range transport of air pollutants project: a technical description, LRTAP-79-5 Atmospheric Environment Service.

Ottar, B. (1978). An assessment of the OECD study on long-range transport of air pollutants (LRTAP). Atmos. Environment, 12, 445-454.

Pasquill, F. (1974). Atmospheric diffusion, New York, John Wiley, 2nd edition.

Powell, D.C., McNaughton, D.J., Wendell, L.L. and Drake, R.L. (1978). A variable trajectory model for regional assessment of air pollution from sulphur compounds. PNL-2734, Battelle, Pacific Northwest Laboratory, Richland, WA, U.S.A.

Prahm, L.P. and Christensen, O. (1977). Long-range transmission of pollutants simulated by a two-dimensional pseudospectral dispersion model. Journ. Appl. Meteor., 16, 896-910.

Prahm, L.P. and Berkowicz, R. (1977). Pseudo-spectral dispersion modelling. Procs. of the 8th ITM, NATO/CCMS, Louvain-la-Neuve, 20-23 September.

Purnell, D.K. (1976). Solution of the advective equation by upstream interpolation with a cubic spline. Mon. Wea. Rev. 104, 42-48.

Reynolds, S.D, Tesche, T.W. and Reid, L.E. (1979). An introduction to the SAI airshed model and its usage. EF 78-53R4-EF 79-31, Systems Applications, Inc., San Rafael, California, U.S.A.

Richtmyer, R.D. and Morton, K.D. (1967). Difference methods for critical value problems, New York, Interscience Publishers, 2nd edition.

Rodhe, H. and Grandell, J. (1972). On the removal time of aerosol particles from the atmosphere by precipitation scavenging. Tellus, 24, 442-454.

Sehmel, G.A. (1980). Particle and gas dry deposition: a review. Atm. Environment 14, 983-1012.

Shannon, J.D. (1981). A Model of regional long-term average sulphur atmospheric pollution, surface removal, and net horizontal flux. Atmos. Environment 15, 639-702.

Sherman, C.A. (1978). A mass-consistent model for wind fields over complex terrain, Journ. Appl. Meteor. 17, 312-319.

Shir, C.C. (1973). A preliminary numerical study of atmospheric turbulent flows in the idealized planetary boundary layer. J. Atmos. Sci., 30, 1327.

Sklarew, R.C., Fabrick, A.J. and Prager, J.E. (1971). A particle in cell method for numerical solution of the atmospheric diffusion equation and applications to air pollution problems. Systems Science and Software report 3SR-844, vol. 1, La Jolla, California, 163 p.

Smith, F.B. (1980). Probability prediction of the wet deposition of airborne pollution. Procs. of the XIth ITM, NATO/CCMS, Amsterdam, 24 - 27 November, p. 1 - 28.

Strang, G. and Fix, G.J. (1973). An analysis of the finite element method. Prentice-Hall International, London, p. 306.

Sutton, W.G.L. (1943). On the equation of diffusion in a turbulent medium, Proc. Roy. Soc., A 182, 48.

Thom, A.S. (1972). Momentum, mass and heat exchange of vegetation. Qu. J. Roy. Met. Soc., 98, 124-134.

Venkatram, A., Ley, B.E. and Wong, S.Y. (1982). A statistical model to estimate long-term concentrations of pollutants associated with long-range transport and its application to emissions from the Sudbury region. Atmos. Environment 16, 249-258.

Veltischeva, N. (1979). A model for evaluation of trans-boundary sulphur dioxide flux with the inclusion of vertical motions.

Voldner, E.C., Olson, M.P., Oikawa, K. and Loiselle, M. (1981). Comparison between measured and computed concentrations of sulphur components in Eastern North America, to appear in Journal of Geophysical Research Proceedings of CAGP Symposium on trace gases and aerosols, C6, 5339-5346.

Walton, J.J. (1973). Scale-Dependent Diffusion. J. Appl. Meteor. 12, 3. WMO symposium on long-range transport of pollutants, Sofia, 1-5 October. WMO report no. 538, p. 284-295.

Wesely, M.L. and Hicks, B.B. (1977). Some factors that affect the deposition rates of sulphur dioxide and similar gases on vegetation. J. Air Pollution Control Association, 27, 1110.

Wyngaard, J.C. (1975). Modelling the planetary boundary layer, extension to the stable case. Bound. Layer Meteor. 9, 441-460.

Zilitinkevich, S.S. (1972). On the determination of the height of the Ekman boundary layer. Bound. Layer Meteor., 3, 141-145.

CHAPTER 3

REMOVAL AND TRANSFORMATION PROCESSES IN THE ATMOSPHERE WITH RESPECT TO SO_2 AND NO_x

R.M. van Aalst and H.S.M.A. Diederen

TNO Division Technology for Society
Delft, The Netherlands

SUMMARY

Mechanisms operative in the removal of air pollutants from the atmosphere are discussed. Formalisms for parameterising dry and wet deposition and chemical conversion are presented. On the basis of literature data, a survey is given on the available information with respect to parameters for the removal of sulphur dioxide, nitrogen oxides and sulphate and nitrate aerosols by these processes. The removal of SO_2 and NO_x in plumes is treated as a separate case.

I. INTRODUCTION

For the development and evaluation of a quantitative model for assessing the effects of emissions of air pollutants on air quality, it is necessary to know, besides the amounts of emitted pollutants and their dispersion, the rate at which they are removed from the atmosphere. Removal is mainly effected by dry and wet deposition and by chemical conversion. A literature study was carried out to determine the contribution of these processes to the removal of some important pollutants, in particular SO_2 and NO_x.

II. DEPOSITION

II.a. Introduction

All substances that are emitted to the atmosphere by human activities will reside in it for a shorter or longer time. These substances can disappear from the atmosphere by chemical conversion in the gas phase, in atmospheric water droplets or on particulate matter and by deposition upon the earth's surface. One distinguishes between dry and wet deposition. Dry deposition takes place by adherence or physical adsorption followed by chemical reactions and/or biological uptake. Wet deposition includes all processes of removal by cloud and rain water and other hydrometeors.

II.b. Dry Deposition

II.b.1. General description. In order to make it possible to account for the effect of dry deposition in model calculations in general, a very simple formula for dry deposition is introduced:

$$F_c = v_d\, C$$

F_c = vertical flux of a pollutant in the surface boundary layer ($\mu g.m^{-2}.s^{-1}$)
v_d = deposition velocity ($m.s^{-1}$)
C = concentration of the pollutant measured at a reference height ($\mu g.m^{-3}$)

v_d, however, strongly depends on atmospheric conditions and the ease of uptake of the pollutant by the surface material. A convenient way to describe the deposition velocity is to introduce the resistance to transfer for the different layers (Garland, 1977):

$$v_d = \frac{1}{r_a + r_b + r_s}$$

r_a = aerodynamic resistance in the boundary layer
r_b = quasi-laminar surface layer resistance
r_s = surface resistance

The resistances r_a and r_b describe the transfer of the pollutant across the air above the surface and are determined by turbulence. The resistance r_b is also dependent on the surface structure and the diffusion coefficient of the pollutant. The surface resistance r_s is dependent on the pollutant and is controlled by the nature of the surface. For the case of vegetation, the term canopy resistance is often used instead.

This canopy resistance can be specified in more detail in terms of stomatal, sub-stomatal and leaf cuticle resistance (Fowler and Unsworth, 1979; Fowler, 1978). The resistance analogy is complicated in the case of deposition of particles, due to the sedimentation of particles under the influence of gravity. In this case the deposition velocity may be approximated by (Slinn and Slinn, 1980):

$$v_d \cong \frac{1}{r_a + r_b + r_a r_b v} + v$$

where v is the particle sedimentation velocity.

Since the deposition is dependent on the physical, chemical and biological characteristics of the surface (e.g. roughness, wetness, degree of acidity, type and status of vegetation etc.) the resistance to transfer should in principle be determined for each type of surface considered in a modelling study. Moreover, the deposition may show considerable variation with time due to changes in meteorological variables as well as surface characteristics. In modelling studies, average resistances or average deposition velocities are often used. It should be noted that this gives only a rough approximation to the real deposition, which is a complicated process.

In some model calculations it is assumed that the pollutant concerned is evenly distributed across the mixing layer. Then the decay rate as a consequence of dry deposition is determined by:

$\lambda_d = \frac{v_d}{H}$, where H is the height of the mixing layer.

In more sophisticated models, the deposition is calculated from the concentration at a reference height (usually 1-10 m), below which the flux is assumed to be independent of height, and determined by deposition alone. The deposition velocity is dependent on this reference height, especially for pollutants with low surface resistance, and under stable atmospheric conditions.

Many reviews on dry deposition have been published. Schwela (1977) discusses general principles, measuring methods and measuring results for several gases. McMahon and Denison (1979) give empirical parameters for dry and wet deposition for many gases and aerosol components. Sehmel (1980) reviews published deposition parameters for gases and particles, and gives model predictions for particle deposition. Hicks et al. (1980) review measuring methods. Flothmann et al. (1982a, b) discuss many aspects of gas and particle deposition to different surfaces and measuring methods. Van Aalst (1982) reviews NO_x dry deposition data, Garland (1983) those for acidic species and ozone, and Fowler (1984) those for SO_2, NO_x, sulphate and nitrate. Hicks and Slinn (1983) give extensive theoretical and experimental considerations with regard to particle deposition. Critical evaluations of theoretical and practical problems can be found in Slinn (1983a) and Slinn (1983b). Many recent results can be found in the proceedings of a conference held in 1982 in Santa Monica (Pruppacher et al., 1983), with a keynote address by Chamberlain (1983). An extensive review of gas and particle dry deposition was also given by the Committee on Atmospheric Transport and Chemical Transformation in Acid Precipitation (1983) (CATCTAP).

II.b.2. Measuring methods. We give a brief survey only of measuring methods. A review on measuring methods for dry deposition is given by Hicks et al. (1980). Details on conditions to be obeyed, data reduction and error analysis can be found in Schwela (1977), Flothmann et al. (1982a) and Hicks and Slinn (1983). The methods can be grouped according to what is measured as follows:

A. measurement of the deposition flux in the air above the surface
B. measurement of the amount deposited
C. measurement of depletion of concentrations in air

It should be noted that the calculation of the deposition velocity from these measurements, and even the determination of fluxes by methods most frequently used, is impossible without the use of models with variable complexity (e.g. models for wind profiles, turbulent transfer, dispersion etc.). Therefore, the determination - especially in field studies - of the deposition velocity (which is a model concept itself!) may be subject to errors due to imperfection in these models alone.

A. Measurement of the deposition flux in the air

If the flux across a horizontal plane not too high above the surface is measured in homogeneous terrain, far enough from sources, the flux is equal to the deposition flux.

From the flux F_c and measurements of the concentration the deposition velocity can be derived from the relationship:

$$F_c = v_d C$$

or from more sophisticated relationships.
The flux of particles by sedimentation cannot be measured by methods of this type currently in use.

Flux measurements can be performed by:

a. The eddy correlation method

In this method the turbulent flux is measured as the average of the instantaneous product of the fluctuating components of vertical wind speed w and concentration c.

$$F_c = \overline{c'w'}$$

Since many years the method has been applied to the measurement of fluxes of heat, momentum and water vapour. In the last decade, the method has been used for flux measurements in the field of CO_2, SO_2, NO_x, O_3 and aerosol particles (see for instance Pruppacher et al., 1983). The most demanding requirement for the technique is that a concentration sensor is needed with a response time of the order of the measurement height divided by wind velocity, typically a second or less. The method is applicable in the field.

b. Gradient and profile methods

In these methods, discussed by Schwela (1977) and Flothmann et al. (1982a), the flux is determined from measurements of concentration C, wind speed and temperature as a function of height z.

The relation used is

$$F_c = K_z \frac{dC}{dz}$$

K_z is estimated from the wind and temperature profile by using theoretical relations. This may, however, introduce some uncertainty in non-neutral conditions of atmospheric stability. In general concentration measurements of high accuracy are needed. The method is frequently used in the field. It can be applied in homogeneous terrain only.

c. Flow reactor measurements

This laboratory technique has been applied to SO_2 and NO_x by Judeikis and Stewart (1976) and Judeikis and Wren (1978). The concentration profile in a tubular reactor with laminar flow is measured; from the concentration and its gradient close to the surface under study the flux is calculated. The resulting deposition velocity is a maximum value that corresponds to values in ambient air for the case that the aerodynamic resistance is negligible.

B. Measurement of the amount deposited

From simultaneous measurements of the amount deposited and of the concentration at a reference height, the deposition velocity may be obtained.

a. Measurements by collection

Many different ways of dust collection (on spheres, plates, filter paper, trays, buckets or in rain water collectors) and subsequent analysis fall in this category. Although this type of field measurement may give an indication of sedimentation of particulate matter and metals (Davidson et al., 1983; 1985), the sampling method can hardly be considered as representative for natural surfaces. Water may be an exception. From several studies, wide variation of the deposition as derived from different types of collectors is reported (Dasch, 1983; Sickles et al., 1983). If natural surfaces such as leaves are analysed, contamination by other mechanisms than dry deposition (e.g. leaching) may interfere (Lindberg and Lovett, 1983).

b. Measurements by tracer methods

When exposing surfaces to known concentration of radioactive pollutants specific deposition measurements may be obtained. This method has been applied in the field as well as in wind tunnels (see Garland, 1978; Flothmann et al., 1982b)

C. Measurements of concentration depletion in air

a. Experiments in enclosures

In these experiments, the concentration in an enclosure placed over the surface under study is measured as a function of time. The deposition velocity is calculated from the apparent first order decay constant K and the ratio of the enclosure volume V to the exposed surface S .

$$v_d = \frac{V}{S} \cdot K$$

In a variant, the deposition is determined from the depletion of the pollutant in air continuously flowing through the enclosure. If the contents of the enclosure are stirred to the extent that the turbulent resistance is negligible, upper limit values of the deposition velocity may be obtained. The method has been applied in the field (see e.g. Böttger et al., 1980; Milne et al., 1979; Johansson et al., 1983). A disadvantage is that vegetation on the surface may be disturbed by conditions of high temperature and humidity often present in these enclosures.

b. Dispersion methods

In these methods described by Schwela (1977) concentrations measured in well-defined situations, for instance directly downwind of a well-characterized isolated source, are compared with values obtained from model calculations with the effects of deposition included. Considerable errors may arise from insufficient knowledge of dispersion and transformation processes and inadequate modelling assumptions. More reliable results can be obtained by using tracers.

By releasing known amounts of an appropriate inert non-depositing tracer from sources of an inert pollutant with known emission strength and measuring tracer and pollutant concentrations as a function of distance, estimates of the removal of the pollutant can be obtained (Horst et al., 1983a). The method has been applied for aerosol particles (Sehmel, 1983).

II.b.3. Dry deposition of sulphur dioxide. The literature on dry deposition of SO_2 is now quite extended. Almost all of the data up to March 1978 have been compiled by McMahon and Denison (1979). A good review on the subject was given at about the same time by Garland (1978). In both articles, values of the surface resistances are listed as well. Rather than repeating this extensive information we give in Table 1, taken from the review by McMahon and Denison, the conclusions drawn at the International Symposium on Sulphur in the Atmosphere (ISSA, 1978). Since 1978, reviews on SO_2 dry deposition have been published by Sehmel (1980), Flothmann et al. (1982b) and Garland (1983) and Committee on Atmospheric Transport and Chemical Transformation in Acid Precipitation (1983). In this section, we will mention some new studies referenced in these reviews or published afterwards.

Table 1. Dry deposition velocities for SO_2 (ISSA, 1978).

Surface	Deposition velocity ($cm.s^{-1}$) typical	range	Comment
dry snow	0.1		if wet, behaves like water
water	0.7	(0.2-1.5)	
calcareous soil	0.8	(0.3-1.0)	wet or dry
acid soil	0.4	(0.1-0.5)	dry
acid soil	0.6	(0.1-0.8)	wet
short grass (0.1 m)	0.5	(0.1-0.8)	
medium crop (1 m)	0.7	(0.2-1.5)	
forest	uncertain	(0.2-2)	

For snow, some new measurements confirm the large variations of the deposition velocity dependent on the state of the snow. Kessler (1979) found $v_d = 2.3\ cm.s^{-1}$ for wet snow and $v_d = 0.64\ cm.s^{-1}$ for frozen snow, from his gradient measurements. Barrie and Walmsley (1978) derived $v_d = 0.25 \pm 0.2\ cm.s^{-1}$ from measurements of accumulation of SO_2 on partially damp snow. Granat and Johansson (1983) found $v_d = 0.1\ cm.s^{-1}$ for dry snow from their enclosure measurements but complete absorption of SO_2 by melting snow. The surface resistance for wet snow is probably negligible, while for dry snow the value is of the order of $1\ s.cm^{-1}$ (Flothmann et al., 1982a).

New field data for dry deposition of SO_2 to water were not found. Water with pH = 8 is considered to be a perfect sink (r_s < 1 $s.cm^{-1}$); for pH > 4, the process is determined by air resistance alone (Garland, 1978; Schwela, 1977). The deposition velocity for SO_2 over water is a relatively well-known function of friction velocity and of the temperature difference between air and water (see Flothmann et al., 1982a); note that over sea, atmospheric stability - as governed by this temperature difference - has a seasonal rather than a diurnal variation. In winter, deposition from continental air masses is enhanced over sea, in summer it is reduced. No new data were found either for soil or building materials. A good discussion of the available data can be found in Schwela (1977). The surface resistance for soil increases by a factor 5-10 when pH decreases from 7.6 to 4.5 and it decreases by a factor 1.6-3 if relative humidity increases from 50-80% (Payrissat and Beilke, 1975).

For building materials with generally high pH values, surface resistance is probably low, typically about 0.5 $s.cm^{-1}$ on a material area basis. The finite capacity for uptake as found in the laboratory study by Judeikis and Stewart (1976), ca. 0.4-3 g SO_2/m^2, is not likely to limit SO_2 uptake in ambient air, where neutralizing components such as ammonia are also present.

Several new studies for grass have been published in recent years. Davies and Mitchell (1983) calculated surface resistance values from a rather large number of gradient measurements as given in Table 2.

Table 2. Surface resistance ($s.cm^{-1}$) for dry deposition of SO_2 to grass (4-20 cm height) in England (Davies and Mitchell, 1983).

		Dry grass	Wet grass
summer,	day	0.70	0.08
	night	1.65	0.30
winter,	day	1.65	0.55
	night	1.30	1.10

The surface resistance for deposition to wet grass is markedly reduced with respect to dry grass. Another aspect is the watering of the grass. Milne et al. (1979) found from enclosure studies in the field v_d = 0.06-0.23 for grass, 15-50 cm high on non-calcareous soil in an arid region, while for well-watered grass (4 cm height) v_d = 0.33-0.58 was found. Hicks et al. (1983) report v_d = 0.18-1.05 (average 0.64 $cm.s^{-1}$) from an eddy correlation study over grassy weeds. Davis and Wright (1985) measured gradients over grass (30-50 cm height) at low SO_2 concentrations and calculated deposition velocities from 4±3 $cm.s^{-1}$ at midday to <0.001 during the late afternoon. Onderdelinden et al. (1984) determined deposition velocities as a function of stability class and wind direction from continuous measurements of SO_2 at 4 and 100 m height on a meteorological tower in the Netherlands during five years. Most of the surroundings were grass. At neutral conditions (class D wind speed 4.5 $m.s^{-1}$) the average deposition velocity varied from 0.4-1.1 $cm.s^{-1}$ depending on wind direction. Winds from south to west showed maximum values, probably due to the increased rain frequency. Annual average v_d was 0.75 ± 0.05 $cm.s^{-1}$.

New information on dry deposition of SO_2 to crop has been obtained from the exposure chamber experiments by Taylor et al. (1983). They found deposition velocities on leaf area basis for bush bean, soy bean and tomato of 0.32, 0.42 and 0.27 $cm.s^{-1}$, respectively. From measurements in light and in the dark, the stomatal and cuticular components of the resistance were estimated. For soy bean or bush bean crop with leaf area index 8 (leaf surface area per unit ground area), the stomatal resistance can be calculated as r_{st} = 0.45 $s.cm^{-1}$, whereas r_{cut} = 2 $s.cm^{-1}$ for bush bean and 1 $s.cm^{-1}$ for soy bean. This may be compared to r_{st} = 1 $s.cm^{-1}$ and r_{cut} = 2.5 $s.cm^{-1}$ as estimated from gradient measurements over a wheat crop in summer as reported by Fowler and Unsworth (1979). These authors discuss the large diurnal variation of the deposition velocity to dry plants due to the strong increase of stomatal resistance during the night. For wet vegetation, the canopy resistance may be lower than dry cuticular resistance. Scenescent crops show increased stomatal resistance. A careful discussion of these and other measurements over wheat and grass by Flothmann et al. (1982a) shows that it is still very hard to explain quantitatively the variation of deposition velocity obtained in field measurements from these factors. In a field fumigation experiment, Sprugel and Miller (1979) estimated v_d = 0.5-0.9 from measurements of the S content of soy bean plants.

Similar leaf area deposition velocities as those for crop measured by Taylor et al. (1983) may be calculated from data reported by Roberts (1974) for woody plants from experiments in an exposure chamber. For white birch, firethorn, white ash and azalea deposition velocities were 0.35-0.23 $cm.s^{-1}$ on a leaf area basis, or 2.8-1.8 $cm.s^{-1}$ assuming leaf area index 8, all at the lowest concentration (0.2 ppm SO_2). The deposition velocity decreased strongly with increasing concentration. This study, which seems to have been overlooked by McMahon and Denison, forms one of the scarce pieces of information about deciduous trees available.

Nestlen et al. (1982) measured deposition velocities of SO_2 by eddy correlation over a rather inhomogeneous area on a tower 50 m high. During 24 hours with steady winds from a sector with deciduous forest, a 24-hour average deposition velocity of 0.45 $cm.s^{-1}$ was obtained; however, the values ranged from +2.5 to -2.5 and advection from nearby sources cannot be excluded. More data for coniferous forest have been obtained in recent years. Fowler and Cape (1983) measured SO_2 deposition to scots pine by eddy correlation. They observed a clear diurnal variation, indicating stomata as a major sink for SO_2. However, an additional mesophyllic resistance was surmised to be present. The deposition velocity ranged from 0.05-1 $cm.s^{-1}$. Rain increased the low night-time deposition velocity from 0.1 to 0.3 $cm.s^{-1}$. Strong diurnal variation was also found by Johansson et al. (1983) in their enclosure experiments with coniferous trees and branches in Sweden. They estimate v_d = 0.5 $cm.s^{-1}$ for summer and v_d = 0.1 $cm.s^{-1}$ in winter. (See also Granat and Johansson, 1983). Dry deposition to wet forest was found to be slow, probably due to the low pH of the water on the needles and branches in Sweden. Galbally and Garland (1979) estimated $v_d \cong 0.2$ $cm.s^{-1}$ from eddy correlation measurements of total sulphur (mostly SO_2) above scots pine forest. Much larger values of the deposition velocity of SO_2 or sulphur may be estimated from studies of rain and throughfall in forests in Germany and the Netherlands. Gravenhorst et al. (1983) estimate v_d = 0.9 $cm.s^{-1}$ in summer and 2.8 $cm.s^{-1}$ in winter. Van Breemen et al. (1982) found in an oak/birch forest and in a scots pine forest in the Netherlands sulphur fluxes in throughfall and stemflow of 1860 mol/ha/year, while about 560 mol/ha/year only is put in by rain. At typical concentrations of 25 $\mu g/m^3$ for SO_2 and 12 $\mu g/m^3$ for sulphate in the Netherlands (Van Aalst et al., 1983b) the deposition velocity for total sulphur would be 1 $cm.s^{-1}$.

In Table 3, an indication is given for the value of the surface resistance for dry deposition of SO_2 to different surfaces. There is still considerable uncertainty in these values.

II.b.4. Dry deposition of nitrogen oxides. Dry deposition of NO_x has been reviewed by Van Aalst (1982). Garland (1983) and Fowler (1984) give more recent results.

Most investigators report low values of dry deposition velocities for NO, often much lower than for NO_2. Deposition velocities to water and sea water, as derived from measurements in enclosures, are of the order of 10^{-3} $cm.s^{-1}$. (Böttger et al., 1980; Gravenhorst and Böttger, 1982). Results for soil are rather variable. Judeikis and Wren (1978) report v_d = 0.13-0.21 to soil and cement surfaces as measured in a flow reactor. However, Gravenhorst and Böttger (1982), using enclosure techniques in the field found v_d < 0.01 $cm.s^{-1}$ for aged cement. For soil, absorption as well as emission of NO was found, probably dependent on the NO concentration. Forest floors, pastures, meadows and lawns were found to release NO at a rate of usually 1-10 μg $N/m^2.s$. In an earlier laboratory study by enclosure

Table 3. Indicative estimates of the surface resistance for dry deposition of SO_2 to different surfaces.

	r_s ($s.cm^{-1}$)	Remarks
snow, dry	> 1	
wet	< .5	
water, pH > 4	< 1	
soil, pH = 7.6 dry	0.4	
wet	0.25	
pH = 4.5 dry	4.0	
wet	1.2	
building materials	0.5	
grass, summer day, dry	0.7	
wet	0.1	
winter day, dry	1.7	higher in
wet	0.6	arid regions
night, dry	1.5	
wet	0.7	
crop, day	0.4-0.7	less, if wet
night	1 -2.5	
trees (coniferous), summer, day	2.	
night	5.-10.	or much lower
winter	> 10	
(deciduous)		not known

techniques, Böttger et al. (1980) found deposition velocities of NO to forest soil and dry or moist sand to be less than 10^{-5} $cm.s^{-1}$. Van Aalst et al. (1983a) reported v_d < 0.01 for clay soil, as measured by the enclosure technique. NO release was sometimes noted, especially after wetting the soil. For grass, growing on soil, a deposition velocity of ca. 0.03 $cm.s^{-1}$ was measured.

From measurements in an exposure chamber by Hill (1971) a deposition velocity for NO to alfalfa of ca. 0.1 $cm.s^{-1}$ can be derived. Law and Mansfield (1982) report uptake rates of NO to sweet pepper leaf from which a value of 0.05 $cm.s^{-1}$ (on leaf area basis) can be derived (cf. Hills value of ca. 0.01 $cm.s^{-1}$ on leaf area basis).

Bengtson et al. (1981) studied dry deposition of NO and NO_2 to scots pine trees and tree branches in chambres and enclosures. In summer, a deposition velocity for NO of 0.05 $cm.s^{-1}$ was estimated. Granat and Johansson (1983) using similar techniques estimate the deposition velocity to spruce and pine in winter to be less than 0.1 $cm.s^{-1}$, and to snow less than 0.03 $cm.s^{-1}$. In Table 4 the reported results are summarized. The uncertainty in the data for soils and the scarcity of data for vegetation prevent definitive conclusions.

Table 4. Reported deposition velocities for NO.

Surface	Deposition velocity ($cm.s^{-1}$)	Remarks
snow	< 0.03	
water, sea water	10^{-3}	
soil	<0.01-0.2	may release NO
meadows, grass	0.03	may release NO
crop	~ 0.1	0.01-0.05 on leaf area basis
conifer tree	0.05	summer
	< 0.1	winter

NO_2 is absorbed by water, soils and vegetation more readily than NO. However, uptake by water is still slow. From an analysis by Lee and Schwartz (1981) it is concluded that the solubility of NO_2 in water and the rate of subsequent reactions in the water is too low to produce deposition velocities of NO_2 to water higher than 0.1 $cm.s^{-1}$. Indeed, Gravenhorst and Böttger (1982) and Böttger et al. (1980) found values of 0.01-0.02 $cm.s^{-1}$ for water and sea-water, using the enclosure technique; Van Aalst et al. (1983a) found $v_d \leqq 0.03$ $cm.s^{-1}$ for sea-water and $v_d \leqq 0.04$ for water. For snow, Gravenhorst and Böttger (1982) and Granat and Johansson (1983) report $v_d < 0.03$ $cm.s^{-1}$.

Judeikis and Wren (1978) found with their flow reactor technique $v_d = 0.77$ $cm.s^{-1}$ for clay soil (pH = 7.9) and $v_d = 0.60$ $cm.s^{-1}$ for sandy loam soil (pH = 8.2). They report re-emission of NO upon prolonged deposition of NO_2 to soil. For cement a deposition velocity of 0.32 $cm.s^{-1}$ was found. However, Gravenhorst and Böttger (1982) report much lower values (0.03-0.1 $cm.s^{-1}$) for aged cement. Heidema and Verhagen (1983), using enclosure techniques, found a strong increase of the deposition of NO_2 to bricks and concrete tiles with humidity. The deposition velocity for brick increased from 0.014 to 0.06 $cm.s^{-1}$ when the water content of the brick was increased from 0.1 to 10% by weight. The deposition velocity to concrete tiles was 0.003 $cm.s^{-1}$ at 15% relative humidity of the air, but 0.1 $cm.s^{-1}$ at near 100%. In contrast, Judeikis and Wren (1978) found no dependence on relative humidity for deposition on soil and cement.

Laboratory measurements with the enclosure technique by Böttger et al. (1980) show lower values for moist sand ($v_d = 0.28$ $cm.s^{-1}$) than for dry sand ($v_d = 0.6$ $cm.s^{-1}$). For a forest soil, $v_d = 0.3$ $cm.s^{-1}$ was found. However, in a similar study in the field (Gravenhorst and Böttger, 1982), deposition velocities for soil were <0.01-0.14

cm.s^{-1}) and for forest soil less than 0.04 cm.s^{-1}. The authors hypothesize that soils handled in the laboratory show increased deposition owing to the enlargement of the surface area. Van Aalst et al. (1983a) found v_d = 0.1 cm.s^{-1} for clay soil in the laboratory.

Deposition velocities to meadows and pastures as measured by Gravenhorst and Böttger (1982) show variable values from less than 0.1 cm.s^{-1} to 0.4 cm.s^{-1} in summer. In winter, measurements on a lawn show values from 0.02-0.25 cm.s^{-1}. Van Aalst et al. (1983a) estimated v_d = 0.35 cm.s^{-1} from experiments with grass in stirred enclosures. Some field measurements of NO_2 deposition to grass (e.g. Wesely et al., 1982) will be considered separately as NO_x-measurements.

Hill (1971) reports very high deposition velocities for NO_2 to oat and alfalfa in his exposure chamber. The deposition velocity for oat during daylight hours was ca. 1.9 cm.s^{-1}, but decreased to 20-35% of this value during the evening. For alfalfa during daylight conditions, v_d = 1.9 cm.s^{-1} was found. Hill did not account for formation of O_3 by photolysis of NO_2 and subsequent uptake by the vegetation. For uptake of O_3, similar values as for NO_2 were found by Hill.

Rogers et al. (1979) reported deposition of NO_2 to plants measured in exposure chambers. NO_2 photolysis effects were shown to be unimportant. From the data, deposition velocities on a leaf area basis can be calculated. For maize and soy beans under daylight conditions values of 0.31 and 0.45 cm.s^{-1} are obtained; Hill's value for alfalfa was 0.2 cm.s^{-1} on leaf area basis. The uptake increased strongly with light intensity. Law and Mansfield (1982) report uptake of NO_2 by sweet pepper leaves corresponding to v_d = 0.13 cm.s^{-1} (leaf area basis).

Rogers et al. (1979) also measured uptake by white oak branches (v_d = 0.09 cm.s^{-1}) and loblolly pine branches (v_d = 0.46 cm.s^{-1}). The lower value for oak was consistent with a lower diffusivity for water. Grennfelt et al. (1983) report experiments with scots pine shoots in exposure chambers in Sweden. Deposition velocities of 0.07-0.17 cm.s^{-1} were found as summer day averages; at night 0.02-0.08 cm.s^{-1} was found in summer, all on a needle area basis. Assuming a leaf area index of 5, typical deposition velocities of NO_2 to pine forest would be 0.4-0.8 cm.s^{-1} during the day and 0.1-0.4 cm.s^{-1} at night, in summer. Substantially lower values were found in winter. Granat and Johansson (1983) also found low deposition velocities (v_d < 0.1 cm.s^{-1} at leaf area index = 3) for spruce and pine in Sweden from measurements with similar techniques. At this latitude the trees did not show any biological activity (CO_2 uptake or transpiration) during winter, at temperatures from +3 to -6°C. In Table 5 the results are summarized.

Table 5. Reported deposition velocities for NO_2.

Surface	Deposition velocity ($cm.s^{-1}$) ground area	leaf area	Remarks
snow	< 0.03		
water, sea water	0.01-0.04		
soil	< 0.01-0.8		dependent on water content?
brick, concrete, cement	0.003-0.3		dependent on water content?
grass [1])	<0.1-0.4		summer, day
	0.02-0.25		winter
crop	1.9	0.13-0.45	day
	0.4-0.7		night
trees	0.4-0.8	0.07-0.17;0.46	summer, day
	0.1-0.4	0.02-0.08	summer, night
	<0.1	<0.03	winter

Since most of the results in Tables 4 and 5 are obtained in stirred enclosures or flow reactors, the values are indicative for the surface resistance: $v_d \cong r_s^{-1}$. However, for high values of the aerodynamic resistance r_a+r_b, especially in low winds under stable atmospheric conditions, deposition velocities might be lower. This is certainly the case for the results reported for crop by Hill and Rogers indicating deposition velocities of 2-5 $cm.s^{-1}$ (if the leaf area index is 5-10). For this kind of canopies, decreasing concentrations lower in the canopy due to uptake of NO_2 by upper canopy leaves might also reduce the bulk deposition velocity.

In Table 6 the results of some studies of NO_x deposition to vegetation in the field by micrometeorological techniques are summarized. In these studies, chemoluminescent NO_x monitors were used, equipped with catalysts to convert NO_2 to NO. These catalysts may convert other NO_x species such as PAN and HNO_3 as well. The deposition velocity of HNO_3 is probably very high (see below). In view of this interference, data from Kasting (1980), who measured the deposition velocity over tall grass in August in Colorado (v_d = 0.33 - 1.77 $cm.s^{-1}$) are suspect since the HNO_3 concentration (0.3-0.6 ppb) was typically ca. 25% of the NO_x concentration (2-3 ppb).

Wesely et al. (1982) measured NO_x deposition to a soy bean crop in August in Pennsylvania by eddy correlation. The NO_x monitor used responded quantitatively to HNO_3 (Huebert, Wesely and Stedman, 1982). The NO_x concentration was less than 13 ppb; the concentration of HNO_3 was not measured. NO was less than 10% of NO_x. The deposition velocities obtained in this study were about 0.3 cm.s^{-1} (0-0.6 cm.s^{-1}; surface resistance 1.3-3 s.cm^{-1}) during the day and less than 0.1 cm.s^{-1} during the night (surface resistance ca. 15 s.cm^{-1}). Although stomatal control of the canopy resistance was obvious, as in the studies of Hill (1971) and Rogers et al. (1979), an additional possibly mesophyllic resistance of ca. 0.5 s.cm^{-1} was noted, and deposition velocities were much lower.

Delany and Davies (1983) report NO_x gradient measurements over cut grass in rural East Anglia during daytime in the spring. Their instrument responded probably quantitatively to HNO_3, but NO_x concentrations were 10-25 ppb and severe HNO_3 interference is not probable at these levels. The surface resistance determined varied from 0.8-5.6 s.cm^{-1} (r_s = 2.4 ± 1.4 s.cm^{-1}) and the deposition velocities from 0.1-0.6 cm.s^{-1}.

Duyzer et al. (1983) measured gradients of NO and NO_x over grass in the Netherlands in September-November. The measuring instrument had a low-temperature catalytic converter and concentrations of NO_x were high (6-150 ppb), so interference by HNO_3 is not probable. NO was often a substantial fraction of NO_x. Deposition velocities were between 0 and 1.5 cm.s^{-1}; however, upward fluxes, possibly due to advection from local sources, were also measured. Chemical reactions between NO, O_3 and NO_2 were demonstrated to influence vertical concentration gradients of these components, but not those for NO_x.

Table 6. Deposition velocity and surface resistance for NO_x measured in the field.

Surface	Deposition velocity (cm.s^{-1})	Surface resistance (s.cm^{-1})	
soy bean	0.3 (<0.56)	1.3-3	summer, day
	<0.1	15	night
grass	0.1-0.6	0.8-5.6	spring
	0.1-1.5		fall

Few data are available for PAN and for nitric acid. Hill (1971) reports a deposition velocity for PAN to alfalfa of 0.6 cm.s^{-1} from his measurements in a recirculating exposure chamber. In a wind tunnel study by Garland and Penkett (1976) values of v_d = 0.21-0.3 cm.s^{-1} (r_s = 2.9-3.4 s.cm^{-1}) were found for soil, and v_d = 0.14-0.26 (r_s = 3.8-7.0 s.cm^{-1}) for grass. Deposition to water was very slow (v_d < 0.02 cm.s^{-1}).

Bos et al. (1978) reported values for the deposition velocity of PAN to sea water of 0.009 $cm.s^{-1}$, to fresh water of 0.011 $cm.s^{-1}$, to dune sand of 0.032 $cm.s^{-1}$, to agricultural soil of 0.06 $cm.s^{-1}$ and to grass of 0.13 $cm.s^{-1}$.
Huebert (1983) and Huebert and Robert (1985) report high deposition velocities for HNO_3 to long (<38 cm) grass; values obtained from gradient measurements in the field ranged from 1.1-4.9 $cm.s^{-1}$ during summer days. The surface resistance was negligible (r_s = -0.03 ± 0.14 $s.cm^{-1}$). From gradient measurements above deciduous forest, preliminary estimates for v_d of several $cm.s^{-1}$ are given. Earlier gradient measurements above grass by Huebert, reported by Kasting (1980) showed considerably lower v_d values and non-negligible surface resistances with much scatter.

II.b.5. Dry deposition of sulphate and nitrate aerosol. As is evident from recent literature (Hicks and Slinn, 1983; Flothmann et al., 1982b; Committee on Atmospheric Transport and Chemical Transformation, 1983; Garland, 1983) dry deposition of aerosol particles is still a confounding subject. Studies in windtunnels (e.g. Chamberlain, 1967; Clough, 1975; Little and Wiffen, 1977; Sehmel and Sutter, 1974) have shown that the deposition velocity is a function of many variables, the most important of which are particle diameter and particle density, friction velocity and surface structure. Particle diameter influences indirectly many mechanisms of particle dry deposition. For the smallest particles (< 0.1 μm diameter), Brownian diffusion carries the particles across the deposition layer close to the surface, where turbulent transfer is negligible. Since the diffusion coefficient decreases with increasing particle size, the deposition velocity also decreases. The velocity of deposition by this mechanism also increases with wind speed and surface roughness.

For larger particles (> 20 μm), sedimentation under the influence of gravity causes an increase in deposition velocity with the square of particle diameter, and also with particle density. Deposition by this mechanism may exceed the limit set to all other deposition mechanisms by atmospheric turbulence ($v_d(max) \leqq (r_a + r_b)^{-1}$, see § II.b.1). For the intermediate region (0.1-20 μm) interception and impaction also play a role. Interception occurs if particles, following the air stream lines, pass obstacles at a distance less than their radius and stick to the surface. Impaction is brought about if the particles, owing to their inertia, cannot follow air stream lines which rapidly deflect around obstacles or in highly turbulent motion. Deposition by impaction increases with particle size and wind speed, and it is strongly influenced by surface structure. Sehmel (1980), in his well-known semi-empirical model of dry deposition of particles uses roughness length as the only model parameter for surface structure. However, it is known from several experiments that the presence of microstructures such as fibres, needles or leaf hairs may enhance deposition velocities several times above the values as expected from roughness length only (Little and Wiffen, 1977; Wells and Chamberlain, 1967; Wedding et al., 1977). According to the model by Sehmel and Hodgson (1980) (see also Sehmel, 1980), the deposition velocity for particles ranges over several orders of magnitude. It shows a pronounced minimum for particles in the range of 0.1-1 μm diameter and increases strongly with wind velocity (or friction velocity) and roughness length. A serious drawback of this model is that it is based on experiments at very low roughness length ($z_o \leqq$ 0.6 cm); extrapolation to higher roughness length is speculative.

Several theoretical models for deposition on more complicated surfaces such as vegetation have been formulated (Legg and Price, 1980; Slinn, 1982a; Davidson et al., 1982; Backe, 1979a,b; 1984; Wiman and Ågren, 1985). Williams (1982) and Slinn and Slinn (1980) (see also Slinn, 1983b) formulated models for aerosol deposition to water and Ibrahim et al. (1983) for snow. Validation of these models is limited, often due to the lack of valid field data (Slinn, 1982a). However, several of the main features seem to be confirmed in field experiments with monodisperse particles or particles in a narrow size range. Neumann and Den Hartog (1985), using eddy correlation, find evidence for low deposition velocities (v_d < 0.05 cm.s^{-1}) over grass for particles in the size range of 0.1-0.5 μm. Sievering (1982) (see also Sievering and Pueschel, 1982) reports confounding results from gradient measurements over rye and wheat for particles of 0.15-0.30 μm, with sometimes much higher deposition velocities. Eddy correlation measurements on the same site by Sievering (1983) showed an average deposition velocity of 0.05 cm.s^{-1} for particles of 0.09-2.5 μm diameter, excluding a period with high wind (~ 7 m.s^{-1}), when the deposition velocity was 0.7 cm.s^{-1}. Eddy correlation measurements in the range of 0.15-2.5 μm by Katen en Hubbe (1983, 1985) do not give conclusive results. Garland (1983) using tracer techniques reports a deposition velocity of 0.05 cm.s^{-1} for 1.8 μm iron oxide particles and 0.42 cm.s^{-1} for 2.4 μm iron oxide particles over grass. Jonas (1984) found from experiments with monodisperse and polydisperse tagged particles between 0.4-17 μm over grass a strong increase with particle size. Deposition velocities of several cm.s^{-1} to grass were found for Lycopodium spores (32 μm) by Chamberlain (1967). The values from Jonas (1984) and Chamberlain (1967) are lower than those predicted by Sehmel's (1980) model (Flothmann et al., 1982b) and seem to increase more with friction velocity.

Gradient measurements with a condensation nucleus counter for particles > 0.03 μm resulted in $v_d \leqq 0.05$ for grass (Garland and Cox, 1982) in accordance with values expected from windtunnel studies for particles in the range of 0.05-0.2 μm. However, eddy correlation measurements reported for particles between 0.05-0.1 μm above snow, grass and soil, maize, pine forest and deciduous forest in winter

(Wesely and Hicks, 1979) show downward fluxes corresponding to v_d values occasionally much larger than this, and also upward fluxes (Wesely et al., 1977; 1983b; Hicks et al., 1982). These measurements have been carried out with instruments responding to aerosol with a large range of sizes. Eddy correlation measurements over grass with an electronic cascade impactor indicate large and sometimes upward fluxes for particles < 0.3 μm (Wesely et al., 1983a). The few measurements over water are contradictory. Gradient flux measurements above a lake (Sievering, 1981) show rather large deposition velocities for particles between 0.1 and 2 μm. Eddy correlation measurements at an ocean shore, however, showed mainly upward fluxes for particles of 0.1-0.9 μm, probably due to particle production by the sea, as well as to a measuring artifact (Sievering et al., 1982; Schmidt et al., 1983). Ibrahim et al. (1983) report deposition velocities over snow for tagged particles of 0.039 and 0.096 $cm.s^{-1}$ for 0.7 μm particles under stable and unstable conditions, respectively, and 0.10 and 0.16 $cm.s^{-1}$ for particles of 7 μm.

From the strong dependence of the deposition velocity on particle size, it is clear that the size distribution of sulphate or nitrate in the aerosol plays an extremely important role in the deposition velocity of the components. According to Whitby (1978), sulphate in continental aerosol is almost completely (95-100%) comprised in one single lognormal mode. The geometric mass mean diameter is 0.48±0.10 μm (density 1.3 g/cm^3) and the geometric standard deviation is 2.00±0.29 μm. For such an aerosol, the sedimentation velocity (on a mass basis) is 5.10^{-3} $cm.s^{-1}$. However, sulphate in marine aerosol is also (for 30-60%) present in a coarse particle mode with a mass mean diameter of ca. 15 μm and a geometric standard deviation 2.7. The sedimentation velocity for this aerosol is 1.8 $cm.s^{-1}$. So, if at a certain location 5% of the sulphate is from this marine coarse particle mode, the sedimentation velocity would be ca. 0.1 $cm.s^{-1}$, or 20 times that of continental sulphate.

As another example, consider the deposition velocities calculated by Davidson et al. (1985) from measured sulphate mass distribution, with the model of Sehmel and Hodgson (1980). They found v_d = 0.64-0.75 for sulphate with a mass median diameter of 0.6-1 μm, while the model predicts v_d = 0.04-0.07 $cm.s^{-1}$ for monodisperse particles with this size. These examples illustrate the large influence of a small fraction of large particles on the deposition velocity . Possibly, these effects explain in part the widely varying deposition velocities reported for aerosol sulphur or sulphate. Everett et al. (1979) calculated from gradient measurements of aerosol sulphur a deposition velocity of 1.4 ± 0.4 $cm.s^{-1}$, while Garland (1978, 1983) could not find any significant gradients over grass (v_d < 0.1 $cm.s^{-1}$). Davies and Nicholson (1982) found from extensive gradient measurements over crop throughout the year a mean deposition velocity for sulphate of 0.08 $cm.s^{-1}$. Doran and Droppo (1983) found v_d = 0.2-0.4 $cm.s^{-1}$ for a limited number of their gradient measurements over grass, while other measurements did not allow for conclusions about the deposition velocity.

Wesely et al. (1983a; 1985) summarize the results of eddy correlation measurements of aerosol sulphur deposition to soil, grass and trees. Deposition velocities to bare soil and to deciduous forest in winter time appear to be very small ($v_d < 0.1$ $cm.s^{-1}$). For pine forest, deposition velocities of 0.4-1.4 $cm.s^{-1}$ (0.7 $cm.s^{-1}$ average) were measured at day time (Hicks et al., 1982) and for a wheat stubble field 0-0.7 $cm.s^{-1}$ (Hicks et al., 1983).
Over grass, 30-50 cm high, the deposition velocity was shown to be dependent on friction velocity and atmospheric stability according to:

$$v_d = 0.002\ u_* \qquad (L \geqq 0)$$

$$v_d = 0.002\ u_* \left[1 + \left(\frac{-300}{L}\right)^{2/3}\right] \qquad (L < 0)$$

where L is the Monin-Obukhov length (m) and u_* the friction velocity. The averaged daily deposition velocity was 0.22 $cm.s^{-1}$, with a day-to-day variation of more than 50%. In all these micrometeorological measurements, the contribution of sedimentation was not measured. However, measurements of particle deposition on leaves in a deciduous forest indicate a large contribution of sedimentation (Lindberg and Lovett, 1983; 1985). For deposition of sulphate to inert surfaces, $v_d = 0.13$ $cm.s^{-1}$ was found; deposition of sulphate to leaves was 50% higher on an average. Taking into account a leaf area index of ca. 5, this would mean a deposition velocity of up to 1 $cm.s^{-1}$ for sulphate.

Höfken et al. (1983) determined dry deposition of several ions to forest (25-30 m high) from the difference of the flux of wet deposition below the canopy (throughfall precipitation) and the wet deposition measured above or outside the forest. Foliar leaching and absorption were not important for sulphate. The deposition velocity of a number of trace elements as found was a function of the mass mean diameter, showing a minimum around 1 μm. For sulphate, the half-year averaged deposition velocity was 1.1±0.4 $cm.s^{-1}$ for beech in summer and 0.6±0.4 $cm.s^{-1}$ in winter. For spruce, they calculated 1.4±0.5 $cm.s^{-1}$ in summer and 1.3±0.5 in winter.

From these considerations, it may be concluded that no reliable overall estimates for sulphate dry deposition can be made. An estimate for grass could be $v_d = 0.2$ $cm.s^{-1}$ for continental aerosol, and $v_d \cong 2$ $cm.s^{-1}$ for marine sulphate, the fraction of which may be estimated from the concentration of NaCl. For deposition to forest, the deposition velocity is probably of the order of 1 $cm.s^{-1}$.

For particulate nitrate, the situation is much more uncertain, since direct flux measurements are lacking. The data from Höfken et al. (1983) and Lindberg and Lovett (1983, 1985) for nitrate cannot be used for nitrate deposition velocity estimates since nitric acid dry deposition and/or nitrate uptake by leaves may have interfered. Nitrate is often stated to have a larger mass-mean diameter than sulphate (as found by e.g. Höfken et al. (1983)). Measurements of the size distribution of nitrate by Yoshizumi and Hoshi (1985) show the presence of ammonium nitrate with a mass mean diameter around 1 μm and sodium nitrate in particles considerably larger in size. In summer, when much of the ammonium nitrate is volatilized, nitrate mass mean diameters may be larger than in winter. It is therefore likely that the dry deposition velocity of nitrate is larger than for sulphate.

II.c Wet Deposition

II.c.1 General description. Wet deposition or precipitation scavenging can be defined as the composite removal process by which airborne pollutants attach to precipitation elements and thus deposit to the earth's surface. This process includes the attachment to cloud droplets by various physical mechanisms during cloud formation and after mixing with an existing cloud (rain-out) and the attachment to falling hydrometeors like rain droplets or snow flakes (wash-out). The possibility of dissociation and chemical conversion of the pollutant in the aqueous phase complicates the wet deposition process even more. Recent reviews of the theory of detailed wet deposition phenomena can be found in Committee on Atmospheric Transport and Chemical Transformation in Acid Precipitation (1983) and Slinn (1982b). Generally, the model predictions of the wet deposition phenomena are accurate to an order of magnitude of one only. As averaging times increase the predictive capabilities can be improved somewhat (because stochastic phenomena average out to mean values).

Quantitative information about precipitation scavenging obtained through field measurements has uncertainties of its own. Many meteorological parameters and specific pollution characteristics of the air parcel in which the precipitation scavenging process is studied have great influence on the mechanisms and on the amount of the pollutant removed by the wet deposition process. As a result of these influences and also as a result of measuring errors the observed scavenging values show a considerable scatter. The field measurements, however, are important to validate model predictions. They also show the range of the scavenging values that occur under ambient conditions.

In most models for mesoscale and long-range transport of air pollutants and in most measurements of precipitation scavenging values, a simplified approach of the wet deposition process is used. Therefore, we will not discuss in detail the complex microscale mechanisms. Our attention is restricted to the simplified macroscale precipitation scavenging parameters. With regard to model calculations the in-cloud droplet scavenging is often not explicitly taken into account. Occasionally, in simulated wash-out the in-cloud scavenging is introduced by assuming an initial or background contamination for the falling rain drops (Adamowicz, 1979).

The efficiency of wash-out depends upon a large number of parameters, such as precipitation rate and precipitation type, droplet size distribution, pH, temperature, vertical concentration gradient of the pollutant etc. Scavenging of particles is strongly dependent on the particle size. Therefore, it is nearly impossible to find a general formula for wash-out.
Often wet deposition is simply considered as an exponential decay process:

$$C_t = C_o \, e^{-\Lambda t}$$

C_t = atmospheric concentration of pollutant at time t
C_o = atmospheric concentration at time zero
Λ = scavenging coefficient (dimension: t^{-1}).

This approach is reasonable in case of a uniformly distributed concentration of an unreactive pollutant, in the absence of sources of the pollutants and with the precipitation scavenging as the only sink for the pollutant concentration. Values for Λ are estimated from theoretical considerations or from field data. As wash-out may increase less then proportional to the intensity of rain, this effect is accounted for in some studies by introducing:

$$\Lambda = \Lambda_w \, J^n, \quad (n<1)$$

Λ_w = wash-out coefficient
J = precipitation rate of liquid water.

Wet deposition of a pollutant strongly depends on the solubility of the pollutant in water. The scavenging ratio is a term which is used for a number of pragmatic purposes. It embodies all the complex in-cloud and below-cloud pollutant scavenging processes.

$$W = \frac{C_w}{C_a} \qquad \text{(dimensionless)}$$

C_w = mass concentration of pollutant in a collected precipitation sample (dimension [mass.L^{-3}]; same as C_a e.g. $\mu g.m^{-3}$)
C_a = mass concentration of pollutant in the air (dimension [mass.L^{-3}]; same as C_w e.g. $\mu g.m^{-3}$).

If the pollutant is an unreactive gas which is dissolved only physically in the water phase the scavenging ratio at equilibrium is equal to the inversed Henry's Law constant.

The wet deposition of a pollutant can be expressed as:

$$D_w = \bar{C}_a \cdot W \cdot J \cdot \Delta t$$

D_w = wet deposition of pollutant per unit of area during time Δt [mass.L^{-2}]
$\bar{C}_a$ = mean mass concentration in the column of air from the ground to the tip of the rain bearing layer [mass.L^{-3}]
J = precipitation rate [$L.t^{-1}$]

Values of W were empirically derived by Engelmann (1970). A more recent survey of empirical deposition parameters is given by McMahon et al. (1979).

If it is assumed that the pollutant is distributed uniformly in the mixing layer, then for the scavenging coefficient it can be written:

$$\Lambda = \frac{W.J}{H} \qquad [t^{-1}]$$

J = precipitation rate in [$L.t^{-1}$]
H = depth of mixing layer [L]

II.c.2 Measuring methods. Measurements of precipitation scavenging parameters can be carried out in artificial rain chambers (Beilke, 1970). Because of the uncertainties in extrapolating to ambient conditions, this method is particularly suitable to compare the wash-out of the component of interest with a known component (e.g. SO_2). The wash-out is determined indirectly by measuring the concentration decay of the component in the gas phase. Generally measurements of precipitation scavenging parameters are done in ambient air. In most cases the difficulties of these methods are of technical nature. The main problems are the conservation of rain water samples and the detection, without interferences, of low concentration levels of the pollutant of interest in the water phase and in the gas phase. The measuring methods used in the field are quite straightforward. The wash-out coefficient is determined from observations of the decay of the pollutant in the air during precipitation events. This is done sometimes by monitoring the pollutant concentration in the air with short integration times during a precipitation event (Davies, 1983) and sometimes by statistical analysis of a great number of hourly averaged observations of the ambient pollutant concentration and the occurrence of precipitation events (Maul, 1978). The scavenging ratio is determinated by simultaneous detection of the pollutant concentration in a precipitation sample and in the air through which the precipitation is fallen. The averaging times for these concentration determination range from a few minutes (Davies, 1983) to daily averaged values (Barrie and Neustadter, 1983) or longer.

II.c.3 Wet deposition of sulphur dioxide. Almost the entire anthropogenic sulphur emission consists of sulphur dioxide. Although it is presumed generally that the wet removal of sulphur from the atmosphere predominantly occurs in the form of sulphate aerosols, the wet deposition of sulphur in the form of sulphur dioxide cannot be neglected. Theoretical studies to predict the precipitation scavenging of sulphur dioxide are done by several investigators (e.g. Hales and Sutter, 1973; Slinn, 1982b; Hales and Dana, 1979a; Barrie, 1981). A comprehensive review of possible mechanisms that play a role in precipitation scavenging is given in Committee on Atmospheric Transport and Chemical Transformation in Acid Precipitation (1983).

The solubility of sulphur dioxide in water is low. It dissolves reversible but it also dissociates in water. The solubility depends on temperature and strongly on the pH of the water. The presence of other pollutants influencing the dissociation equilibria or the pH (like acid aerosols, nitric acid, ammonia) can have considerable effect on the absorbed amount of SO_2. Consequently the scavenging ratio of SO_2 is not a constant, but rather strongly dependent on rain pH and temperature (Hales and Sutter, 1973; Barrie, 1978; Barrie, 1981). Barrie (1978) has shown that for the range of precipitation pH's commonly observed in the atmosphere almost all the dissolved SO_2 is in the form of bisulphite. Under equilibrium conditions a simplified relation between the scavenging ratio and the temperature and pH of the drop can be derived (Barrie, 1981):

$$\log_{10} W_{SO_2} = \log_{10} (K_1 . K_H) + pH$$

with $K_1 . K_H = 6.22 * 10^{-8} \exp. (4755.5/T)$

(for T ranging from 273 to 303°K)

Because of its low solubility the occurrence of liquid phase oxidation reactions which transform dissolved SO_2 into sulphate can enhance significantly the total amount of SO_2 removed from the atmosphere. Due to the small dimensions of cloud droplets, equilibrium between gas phase and solution is reached within a short time, and as the residence time of cloud droplets is long enough, chemical reactions in the droplets may play an important role. As falling rain drops have much larger sizes and smaller residence times, chemical processes in the liquid phase are less important. Fowler

(1980) estimated that a fraction of ~20% of the sulphate in rain is due to liquid phase oxidation of SO_2, but also higher fractions are found. Barrie and Neustadter (1983) estimated that for eastern Canada possible 70 ± 30% originates from SO_2 oxidation, Scott (1982b) calculated for storm conditions in the northeastern U.S. 40-70%. On the contrary Pena et al. (1982) observed at Central Pennsylvania that the sulphate content in rain was negatively correlated with sulphur dioxide concentration in air at the surface, suggesting that for this area the contribution of the SO_2 oxidation is of minor importance. The amount of SO_2 removed by reactive scavenging will scatter considerably depending on several atmospheric conditions (see also: III. Chemical conversion in the atmosphere). We will further discuss only the unreactive scavenging of SO_2. As already mentioned, due to the small dimensions of cloud droplets, equilibrium between SO_2 in the gas phase and the SO_2 in solution in clouds is reached within a short time. Observations in clouds are in agreement with this (Daum et al., 1983). Under most atmospheric conditions, in the absence of adjacent sources of SO_2, which cause non-uniform concentration profiles, and without oxidation processes in the liquid phase, the wet deposition process of SO_2 can be described as equilibrium scavenging. Sometimes the loss by wet deposition by this process is neglected. However, the fraction of sulphur that is removed is minor but not always insignificant (Hales et al., 1978). Davies (1976) observed in a heavily-industrialized area in Sheffield, U.K., that an averaged fraction of 47% of the rainborne sulphur was rainborne SO_2. Davies (1979) found a fraction of 14% in rural Norfolk, U.K. and a fraction of 22% at an urban site of the city of Norfolk. Hales and Dana (1979a) estimated fractions ranging from ~10-30% for the northeastern U.S. and calculated from the LAPPES project data values ranging from 6-27% for western Pennsylvania. Barrie and Neustadter (1983) estimated a fraction of generally less than 10% for eastern Canada. Högström (1974) reports limited measurements in Uppsala, Sweden, with values ranging from 0 to 13%.

Because of the different meteorological conditions and the specific air pollution characteristics, measurements of precipitation scavenging of SO_2 on a particular place cannot be generalized. In spite of the uncertainties and the scatter in observed values of washout coefficients and scavenging ratio, these measurements are of importance to illustrate the ranges in the ambient atmosphere and a generally coherent picture can be obtained. Garland (1978) reviewed observations of wash-out coefficients of SO_2 of several investigators and concluded that the wash-out coefficients are almost all of the order of $10^{-5}.s^{-1}$.

In the survey of deposition parameters of McMahon (1979) values ranging from $0.4*10^{-5}$ to $6*10^{-5}.s^{-1}$ are reported. Laboratory investigations of Beilke showed a relationship of $\Lambda_{SO_2} = 17*10^{-5} J^{0.6} s^{-1}$ (J = rainfall rate $mm.h^{-1}$). Maul (1978) used hourly measurements of atmospheric sulphur dioxide and rainfall at ten monitoring sites at rural East Midlands, U.K. They analysed the data for non-frontal rain events and calculated $\Lambda = (2.85 \pm 0.29) * 10^{-5} * J.s^{-1}$, where J is the rainfall rate in $mm.h^{-1}$. Davies (1983) calculated wash-out coefficients from measurements of airborne and rainborne SO_2 levels at two heights in an industrialized area (Sheffield, U.K.), assuming a linear change of concentration of atmospheric SO_2. He observed values ranging from $0.01*10^{-5}$ to $8.56*10^{-5} s^{-1}$ with a median value of $2.6*10^{-5} s^{-1}$.

Martin (1984) used the same statistical method as devised by Maul (1978). Martin analysed the data of five years continuous monitoring at the rural site of Bottestford, U.K. The estimate relationship between the wash-out coefficient of SO_2 and the rainfall rate J (in $mm.h^{-1}$) was $\Lambda = (2.61 \pm 0.14) * 10^{-5} * J s^{-1}$. This relationship is consistent with that found by Davies (1983) and Maul (1978) and seems not to conflict with other observations of the SO_2 scavenging coefficient.

Even of more practical applicability for modelling activities is the scavenging ratio of SO_2. From the equation of Barrie (1981) it can be derived that under ambient conditions likely to occur scavenging ratios for SO_2 will fall in the range of 10^3-10^6 (dimensionless, in mass/volume rain divided by mass/volume air; in this section all values of scavenging ratios are converted to dimensionless values ($\mu g.m^{-3}/\mu g.m^{-3}$). In the survey of deposition parameters McMahon (1979) only presents the values for the SO_2 scavenging ratio obtained by Georgii and Beilke. For rain storm with a precipitation amount of 0.3 mm they reported a scavenging ratio of $1.6*10^5$ and for an event of 11-20 mm they reported a value of $1.6*10^4$.

From the curves of the relation between SO_2 concentration in the air and the SO_2 concentration in the rain presented by Davies (1979) estimated values for the scavenging ratio can be derived. For the data of the industrialized site at Sheffield (Davies, 1976) the scavenging ratio is in the order of 10^4, and for the Norfolk rainfall it is about $2*10^4$. Davies (1983) analysed the data of the hourly observations at the industrialized area of Sheffield, already mentioned, and concluded that, although the measurements are lower, they show a reasonable agreement with predicted values, using the relationship of the scavenging ratio and the pH and temperature derived by Barrie (1981). To assess the effect of a vertical gradient, which is likely to occur in the vicinity of sources of SO_2, measurements at two heights were made by Davies (1983). This measurement showed a good correlation between observed and predicted scavenging ratios, although the observed ratios were still lower than the predicted ones. A possible reason for the lower values observed is that in this environment the equilibrium scavenging assumption is invalid. During an intensive study of three individual rain events at the rural site in Norfolk short-period sampling of rain and air concentrations was carried out by Davies (1983). The mean of the predicted ratios, $6.1 * 10^5$, was in agreement with the mean value of the observed ratios, $5.3 * 10^5$.

The measuring results of the scavenging ratio of SO_2 are all in the range of 10^3-10^6, as predicted from the relationship between the scavenging ratio and the pH and the temperature. Near sources, as a consequence of vertical gradients of the SO_2 concentration, equilibrium scavenging may not occur. The detailed observations by Davies (1983) at a rural site with short-period sampling are very close to the predicted values. Therefore it can be concluded that the equation for the simplified relation given by Barrie (1981) can be used for modelling of equilibrium scavenging of SO_2.

II.c.4 Wet deposition of nitrogen oxides. The solubility of NO_2 in water is low. The Henry's Law coefficients for NO and NO_2 are $2*10^{-3}$ $M.atm.^{-1}$ and $1*10^{-2}$ $M.atm.^{-1}$, respectively (Daum et al., 1983). NO_2 dissolves slowly in water with decomposition (Lee and Schwartz, 1981). Observations of ambient concentrations and interstitial concentrations in clouds are consistent with the Henry's Law coefficients of NO and NO_2 (Daum et al., 1983; Schwartz et al., 1983). Measurements of ambient concentrations by several investigators show that even after abundant rain the NO and NO_2 concentrations have hardly decreased. From these observations and because of the virtual absence of excess nitrate or nitrite in precipitation samples collected beneath plumes with high concentrations of nitrogen oxides the conclusion can be drawn that direct uptake of NO and NO_2 by precipitation and cloud droplets is negligibly slow. (Committee on Atmospheric Transport and Chemical Transformation in Acid Precipitation, 1983). Measurements of the wash-out coefficient of NO_2 in an artificial rain chamber are carried out by Beilke (1970). He obtained values of the scavenging coefficient of NO_2 of $4*10^{-5} * J^{0.6}$ s^{-1} (with J in $mm.h^{-1}$) which were four to five times lower than the corresponding values for SO_2 obtained by him with the same method.

Martin (1984) has estimated the below cloud scavenging coefficients (wash-out coefficient) for NO and NO_2 by statistically comparing hourly ground level concentrations just before and at the onset of heavy, non-frontal rain during five years of continuous monitoring at a rural site in the U.K. He estimated values for NO and NO_2 of $1*10^{-5}*J$ s^{-1} and $2.2*10^{-5}*J$ s^{-1}, respectively (J in $mm.h^{-1}$) (40% and 80% of the value of SO_2). The solubilities of NO and NO_2 relative to that of SO_2 are very low. The effective Henry's Law coefficient (with dissociation taken into account) of SO_2 is, even at lower pH values, about four orders greater than the corresponding values for NO or NO_2 (Schwartz et al., 1983). Thus the differences in solubilities may not account for the values observed. As a possible cause for the values observed Martin mentions the occurrence of chemical conversion processes of NO_x in the aqueous phase and the fact that the measuring method used for NO_2 essentially measures the sum of NO_2, nitric acid and nitrate aerosols.

The conversion of NO_x to nitric acid in the gas phase is of great importance for the removal of NO_x by wet deposition processes. Nitric acid is a highly soluble gas and is therefore rapidly scavenged by cloud and rain droplets. Observations of nitric acid concentrations in clouds show low concentrations (below the detection limit) in the interstitial air (Daum et al., 1983). Levine and Schwartz (1982) estimated scavenging coefficients for the removal of nitric acid vapour by both cloud and rain drops from model calculations. They estimated a value of 0.2 s^{-1} for cumulus clouds and recommend values of $\sim 1*10^{-4}$ s^{-1} and $\sim 4*10^{-4}$ s^{-1} for rainfall rates of 1 and 15 $mm.h^{-1}$, respectively for ambient atmosphere.

The scavenging of gases by snow can be ignored generally (Slinn et al., 1978). However, for highly soluble gases like nitric acid scavenging is observed. Huebert et al. (1983) measured the scavenging of nitric acid during eleven snowfall events. They reported a lower limit value of the snow scavenging coefficient of nitric acid vapour of $25*10^{-5}$ s^{-1} and a value of $5*10^3$ for the scavenging ratio of nitric acid (corrected for aerosol nitrate by factor analysis of snowfall data).

Generally it is assumed that the precipitation scavenging of PAN is not important as a removal process of nitrogen compounds. However, quantitative information is scanty. Recent investigations have been made of the Henry's Law constant and the decay rate of PAN in acidic water (Holdren et al., 1984). For the Henry's Law constant at 10°C a value of 5±1 $M.atm^{-1}$ is determined. This value is several orders of magnitude higher than the value for NO_2, slightly higher than the value for SO_2, but several orders of magnitude lower than the effective Henry's Law constant (with dissociation) of SO_2 (Schwartz et al., 1983). The decomposition of PAN in acidic water (pH 4-6) at 5°C is a slow process with a decay rate of $7.0 \pm 0.3*10^{-3}$ min^{-1}. Holdren et al. (1984) conclude that the contribution of PAN to precipitation nitrate levels is probably not important.

II.c.5 Wet deposition of sulphate and nitrate aerosols. In this section we will discuss the wet removal of sulphate and nitrate aerosols. In literature most investigations are carried out in regard of atmospheric aerosols as a whole or of aerosol sulphate. Investigations specific for wet deposition of aerosol nitrate are scanty. However, most conclusions can be generalized to any water-soluble aerosol with the same particle size distribution as the one investigated. The precipitation scavenging of sulphate and nitrate aerosols is a result of nucleation scavenging (the particles act as cloud condensation nuclei) and impaction of particulates by hydrometeors. Most investigators assume that nucleation scavenging is the most important process for particles in the size range of 0.1-5 μm, typical for ambient nitrate and sulphate aerosols (Junge, 1963; Makhon'ko, 1967; Fowler, 1980; Charlson and Rodhe, 1982; Leaitch et al., 1983; Radke, 1983; Committee on Atmospheric Transport and Chemical Transformation in Acid Precipitation, 1983; Hegg et al., 1984b). Theoretical arguments predict that the nucleation scavenging efficiency (the fraction of the aerosol attached by the cloud water because they act as condensation nuclei) of sulphate aerosols lie in the range of 50 to 100% (Junge, 1963; Fisher, 1982). Field measurements confirm these predictions. Scott and Laulainen (1979), considering a single case, observed a sulphate aerosol nucleation scavenging efficiency of 55%; Hegg and Hobbs (1983) found a sulphate aerosol nucleation scavenging efficiency in several types of clouds of (63±53)%; Hegg et al. (1984) measured a nucleation scavenging efficiency of 70% for aerosol in the size range of 0.1-1 μm; Sievering et al. (1984) observed in two in-cloud studies a removal efficiency of ~90% of sulphate aerosol. Measurements by Daum et al. (1983) show low interstitial aerosol concentrations in clouds relative to ambient air suggesting high in-cloud scavenging efficiencies.

Many investigations on the scavenging coefficient of aerosols for in-cloud and below-cloud conditions and for rain as well as snowfall are reported. Makhon'ko (1967) concluded from theoretical work and experimental data that for atmospheric aerosols the in-cloud or rain-out scavenging coefficient has values of 10^{-4} - 10^{-3} s^{-1} and the below-cloud or wash-out scavenging coefficient is 10^{-5} - 10^{-4} s^{-1}. Sievering et al. (1984) estimated an in-cloud scavenging coefficient for aerosol sulphur of 10^{-3} (10^{-4} - 10^{-2}) s^{-1} from one cloud event observation. During rain, according to Garland (1978), sulphate is removed with a time constant of the order of 10^{-4} s^{-1}. McMahon and Denison (1979) give a survey of empirical scavenging coefficients of several investigators up to 1977. The values for the observed scavenging coefficients scatter from 10^{-3} to 10^{-5} s^{-1}. When plotted the values with specified particle size, the values show a satisfactory mutual agreement. This illustrates the importance of the particle size. The great influence of the size distribution of aerosols on the wash-out coefficient is demonstrated by calculations of Dana and Hales (1976). Davenport and Peters (1978) observed wash-out coefficients of particles in two subranges, around 0.01 μm and 1 μm, during three precipitation events. For the two size ranges about the same apparent wash-out coefficients were observed, varying between $2*10^{-5}$ s^{-1} and $1*10^{-4}$ s^{-1} depending on rain drop size distribution and precipitation intensity.

Scott (1982a) presented an estimated scavenging coefficient for soluble particles as a function of precipitation rate for rainfall as well as for snowfall. The coefficient is intended for regional or mesoscale models. For rain he estimated $\Lambda = 3.5*10^{-4} * J^{0.78}$ s^{-1} and for snow $\Lambda = 2.4*10^{-4}*J^{1.0}$ s^{-1}, where J is the precipitation rate (in $mm.h^{-1}$) which varies with the altitude. He estimated an uncertainty in the scavenging coefficients of a factor of 2 or 3 for rain and a factor of 6 for snow. Huebert et al. (1983) measured a value of $5*10^{-5}$ s^{-1} for the scavenging coefficient of sulphate and nitrate aerosols by snowfall.

For modelling purposes also the scavenging ratio of aerosol is important. In this section the scavenging ratios cited are converted to the dimensionless form for uniformity's sake (in $\mu g.m^{-3}/\mu g.m^{-3}$). Generally it is assumed that the scavenging ratio decreases as the precipitation amount increases (Slinn, 1982b). For atmospheric aerosols values of scavenging ratios in the order of 10^5-10^6 are expected (Slinn et al., 1978; Slinn, 1982b). Calculated values, as a result of modelling of only the below-cloud sulphate aerosol scavenging (Carmichael and Reda, 1982), vary between 10^3 and $2*10^4$, depending on the drop size. These lower values suggest an important role of the in-cloud processes. Lindberg (1982) observed for five rain events at a site in the Tennessee Valley values of the sulphate scavenging ratio between $4*10^5$ and $2*10^6$ with a mean value of $8*10^5$. Barrie and Neustadter (1983) calculated sulphate aerosol scavenging ratios from daily mean concentrations observed at four rural locations in eastern Canada during a period of one to two years. They found a range of geometric means for the four locations from $8.1*10^5$ to $1.1*10^6$ with geometric standard deviations of 2-3. Snowfall events seem to result in somewhat lower values than rainfall events. The observed relationship between the scavenging ratio W and the precipitation amount P was

$$W = \text{constant} \;.\; P^{-(0.407\pm0.052)} \text{. where } P = \text{amount of rain in cm.}$$

The mean values observed in Canada are consistent with the mean values of $8.3*10^5$ to $1.3*10^6$ observed at rural locations in Norway (Whelpdale, 1981). Hales and Dana (1979b) give a relationship for the scavenging ratio and the precipitation amount of

$$W=1.9*10^6 \;.\; P^{-0.88}$$

Niemann (1983) analysed the sulphate scavenging ratios of exceptionally wet sulphate episodes (highest wet sulphate depositions) in the MAP3S, the EPRI-UAPSP and the APN network. The scavenging ratios show considerable scatter with values of about $\sim10^5$, $\sim5*10^5$ and $\sim10^6$, respectively, for the three networks mentioned. The scavenging ratios for these exceptional episodes appeared to be independent of precipitation rates.

The scavenging of aerosols by snowfall is considered a more efficient process than scavenging by rainfall (Raymor and Hayes, 1983). We have already mentioned, however, that it was observed that scavenging rates by snowfall are lower than by rainfall as well (Scott, 1982a; Barrie and Neustadter, 1983). Huebert et al. (1983) measured a value of $8*10^5$ for the scavenging ratio by snowfall for both sulphate aerosols and nitrate aerosols. Scott (1981) found that the sulphate scavenging ratio was around $8*10^4$ for unrimed snow, but near $8*10^5$ when the snow was rimed. Thus the specific conditions for snowfall seem to be important. Although the differences between rain and snow scavenging are not totally clear, great differences seem unlikely.

The observations of the scavenging coefficient as well as of the scavenging ratio for sulphate aerosols and nitrate aerosols are rather consistent. For the scavenging coefficient of sulphate and nitrate aerosols an averaged value of 10^{-4} s^{-1} and for the scavenging ratio of sulphate and nitrate aerosols an averaged value of $5*10^5$ seems reasonable. The scatter of these scavenging parameters, however, is considerable. Uncertainties of one order of magnitude have to be taken into account.

III. CHEMICAL CONVERSION IN THE ATMOSPHERE

III.a Introduction

Air pollution chemistry is often rather complicated in the sense that equilibria and higher-order reactions play an important role. Our knowledge of the chemical conversion process stems from fundamental kinetic and mechanistic research in the first place. Chemical kinetic schemes are built from the kinetic parameters of the reactions that have been shown to be important. However, considerable uncertainty can remain in the results of calculations with such schemes due to experimental uncertainty of kinetic parameters. Therefore, the schemes are tested as a whole against smog chamber experiments or in field studies. Limitations of validation by smog chamber experiments are:

- that experiment durations will be generally less than 12 hours, because after such a period wall reactions and wall losses, that are specific for the smog chamber become dominant;
- that the dominant mechanisms by which important reactants like the OH radical are generated in smog chambers differ from dominant generation mechanisms outdoors (Carter et al., 1982).

Validation of chemical schemes by field studies is only possible if a perfectly accurate and complete data base for chemical and meteorological data is available. Still, then, deviation between measured and calculated concentrations may arise from shortcomings in non-chemical assumptions used in the model or from errors in the numerical techniques used. In an analysis of sophisticated air quality simulation models for the prediction of ozone concentrations, Seinfeld et al. (1977) estimated an uncertainty in absolute predictions of about 50% due to uncertainties in the chemical kinetic models. Although recent sophisticated models incorporate updated and further detailed transport and transformation processes, there remain many areas of uncertainty.

Air pollution models can be classified according to the treatment of chemical processes on the one hand and of dispersion and meteorological factors on the other. In most simple dispersion models like, for instance, the Gaussian plume model, chemical conversion can only be specified as a net first-order rate constant. Such an approximation allows for linear combination of the contributions of different sources to local concentrations, thus cutting down computational effort. On the other hand, simple models for photochemical air pollution like box models (Derwent et al., 1979) and the EKMA model [EPA, 1977)] have a chemical reaction scheme as their main component, whereas dispersion and meteorology is dealt with in a rather limited way.

Only few models like the SAI model (Reynolds et al., 1979) treat chemistry as well as dispersion in more detail. A review of operational models is given by Van den Hout and Van Dop (1985) as Chapter 2 of this report of the NATO/CCMS pilot study on air pollution control strategies and impact modelling. Since 1982 the ADOM-TADAP Model development program is carried out initiated by Canadian and German authorities to incorporate the state-of-the-art information on all the relevant physical and chemical processes. This model will be one of the most sophisticated computer simulation model of tropospheric processes.

In the simple as well as in the more sophisticated models with detailed chemistry, the chemical kinetic scheme is an intrinsic part of the model. It is our goal to find effective first-order conversion rate constants. In this document, we will discuss kinetic schemes only to derive these rate constants.

III.b Concentration of Hydroxyl Radicals in Ambient Air

The most important conversion of many compounds is the reaction with hydroxyl radicals (OH). This applies, for example, to CO, SO_2, NO_x and hydrocarbons. Unfortunately only a few determinations of OH radical concentrations have been carried out and some of these are unreliable. Perner et al. (1976) reported UV absorption measurements at Jülich (Federal Republic of Germany) and found values of about 2×10^{-7} ppm. Earlier, in the United States, values were measured by Wang et al. (1975) with the aid of laser fluorescence up to 2.5×10^{-6} ppm in the open air. From later publications of this group (Hanabusa et al., 1977; Selzer et al., 1979) it appeared that interference with other components occurred, so that these results must be considered unreliable. With a laser fluorescence technique, developed by Davis and coworkers (Davis et al., 1979a) values of about 4×10^{-7} ppm were found in the USA nearby a large power station, but outside the plume (Davis et al., 1979b). From measurements based on the ^{14}CO chemical method of Campbell et al. (1979) in the USA in the open air concentrations of 10^{-8} ppm up to 1.5×10^{-7} ppm were deduced.

In Hübler et al. (1984) measurements of OH radical concentrations for 1979 through 1981 at two locations in Germany using long-path UV absorption spectroscopy are reported. The possible spectral interferences are identified and corrected noontime OH concentrations at sunny days averaged to $6 * 10^{-8}$ ppm. The highest concentration observed was $1 * 10^{-7}$ ppm. In this paper also a review of other recent measurements is given. Campbell et al. (1982) measured OH concentrations of $0.7 - 15 * 10^{-7}$ ppm at Washington in 1979/1980 and L. Davis et al. (1982) $0.1 - 1.3 * 10^{-7}$ ppm at Niwot Ridge in October 1981. Recently ambient measurements with a new technique FAGE (fluorescence assay with gas expansion) were made (Hard et al., 1984). The OH concentrations measured are consistent with other measurements.

All these values were measured in the day time. As these values are insufficiently representative, complete and accurate, estimates about the variation of concentration during the day, the season and with latitude can only be derived from model calculations. Calculations of Hov and Isaksen (1979) for clean and urban areas give OH concentrations with day maxima varying from 6×10^{-8} to 4×10^{-7} ppm; the smallest values were obtained during winter. The OH concentration increases with light intensity and is negligible in the absence of sunlight. High concentrations of NO_x tend to lower the day maxima. The influence of temperature is small. Weinstock et al. (1980) deduced from model calculations of the concentration of OH radicals in the troposphere that light intensity and NO_x concentration are the most important factors. In a recent review (Committee on Atmospheric Transport on Chemical Transformation in Acid Precipitation, 1983) these data and the theoretical estimates based on computer models are found reasonably consistent and are accepted as characteristic in a polluted atmosphere.
From light intensity measurements carried out in Delft, the Netherlands, during the years 1973, 1975, 1976 and 1977 (Brunner et al.,

1978) it can be estimated that during the months of June, July and August the integrated UV light intensity corresponds to an averaged value of about 70 $W.m^{-2}$ in 12 hours. During winter this amounts to about 8 $W.m^{-2}$, during autumn and spring to about 20 $W.m^{-2}$. Assuming a maximum value for the summer day maximum of 4 x 10^{-7} ppm, as calculated by Hov and Isaksen (1979), and also assumed by Calvert and Stockwell (1983a), computer simulations show that for these conditions the 24-h average of OH concentration is ~7 * 10^{-8} ppm (Committee on Atmosphere Transport and Chemical Transformation in Acid Precipitation, 1983). This value corresponds with a day-time average OH concentration of ~1.5 * 10^{-7} ppm during summer, 1.7 * 10^{-8} ppm during winter and 4.3 * 10^{-8} ppm during spring and autumn. Supposing that the concentration at night is $\leqq$ 8 * 10^{-9} ppm (Calvert and Stockwell, 1983a) a year-averaged concentration of 3-3.5 * 10^{-8} ppm can be estimated. The data are summarized in Table 7.

Table 7. Estimated averaged concentrations of OH radicals in the Netherlands.

Period	Concentration (10^{-7} ppm)
night	0 (<.1)
day (summer)	1.5
day (autumn, spring)	0.4
day (winter)	0.2
year-averaged	0.3

With the aid of this table and the reaction constants of the reactions with hydroxyl radicals, the conversion of various compounds by this mechanism can be estimated. In view of the limited number of data the uncertainty of the absolute values of these concentrations is considerable.

III.c Conversion of Sulphur Dioxide

III.c.1. Theoretical Studies and Laboratory Measurements. The literature concerning the conversion of sulphur dioxide into sulphate is very comprehensive. Reviews on the different possible mechanisms and laboratory measurements can be found in the proceedings of the International Symposium on Sulphur in the Atmosphere (ISSA, 1978), particularly the papers by Calvert et al. (1978) and Eggleton and Cox (1978) for homogeneous gas phase reactions, and by Beilke and Gravenhorst (1978) and Hegg and Hobbs (1978) for conversion in water droplets. More recent reviews are Calvert and Stockwell (1983b), Committee on Atmospheric Transport and Chemical Transformation in Acid Precipitation (1983) and Atkinson and Lloyd (1984).

These recent reviews agree with regard to the dominant role of the OH radical concerning the gas phase conversion of sulphur dioxide.

The conversion by HO_2 and RO_2 appears to be negligible at atmospheric conditions (reaction rate constants are $<2 * 10^{-3}$ $ppm^{-1}.min^{-1}$ and $<7 * 10^{-2}$ $ppm^{-1}.min^{-1}$, respectively). The conversion by Criegee intermediates, generated by ozone-olefin reactions, is not totally clear but quantitatively it is of minor importance.

The recent conclusion that the reaction SO_2 + OH is not an OH chain terminator but that it generates HO_2 radicals is important. In this way, SO_2 conversion enhances slightly the rate of NO to NO_2 conversion and ozone formation ($NO + HO_2 \rightarrow NO_2 + OH$) (Stockwell and Calvert, 1983, Calvert and Stockwell, 1983b, Meagher and Olszyna, 1984).
Although the detailed reaction mechanism is not totally clear the overall reaction recommended (reaction rate constant: $1.4 * 10^3$ $ppm^{-1}.min^{-1}$) is:

$$HO + SO_2\ (+\ O_2,\ H_2O) \rightarrow H_2SO_4 + HO_2$$

On a sunny summer day in an urban area the maximum conversion rate is estimated at 2-4% h^{-1} at a daily averaged rate of 0.5 - 1% h^{-1}, on the basis of model calculations. In clean air the maximum conversion is about 1 - 2% h^{-1}. At night the conversion rate in the gas phase is low and is possibly determined by reaction between ozone and olefins and NO_3 radicals with formaldehyde.

With respect to the conversion in the liquid water phase (cloud, rain, fog) the situation is less clear. Sulphur dioxide dissolved in droplets can be oxidized by H_2O_2, ozone, radicals such as OH, HO_2 etc. and oxygen catalyzed by metal ions, such as Mn and Fe. Current theoretical understanding of the oxidation rates of dissolved SO_2 by the various proposed mechanisms are shown by Martin (1983). Oxidation rates of SO_2 in liquid phase are of the order of a hundred % hr^{-1}. The lifetimes of individual clouds, however, are short, so that the long-term average oxidation rates in cloudy air and in the gas phase may be of the same order of magnitude (Committee on Atmospheric Transport and Chemical Transformation in Acid Precipitation, 1983).

The relative importance of the liquid phase and gas phase processes for various conditions may vary, depending on a variety of meteorological conditions such as the liquid water content, precipitation, relative humidity, solar radiation and the pH of the droplets and the concentrations ofvarious pollutants. The photochemical reactions of NO_x and hydrocarbons supply the necessary oxidizing agents for the gas-phase conversion as well as for the aqueous-phase conversion of SO_2. In liquid phase oxidation at ambient conditions the rate-determining step is not the transport of sulphur dioxide to and in the droplet but the chemical conversion (Freiberg and Schwartz, 1981; Schwartz and Freiberg, 1981). This conversion pathway would especially occur in clouds, in plumes and in urban fog. Moreover, it is possible that the sulphur dioxide is converted on the surface of particles (Judeikis, 1978; Novakov, 1975; Liberti, 1978).

Recent results of heterogeneous SO_2 conversion on and in particles were presented at the Advanced Study Institute on Chemistry of Multiphase Atmospheric Systems Corfu 1983. Guesten (1983) studied photocatalytic oxidation of SO_2 on coal fly ash, cement dust and soot; Dlugi (1983a) studied aqueous phase oxidation of SO_2 in fly ashes, cements, soots, industrial dusts, volcanic ash, mineral dusts and sea salt; McLaren (1983) examined SO_2 oxidation on activated carbons and coal and oil fly ashes. The SO_2 oxidation is catalyzed most effectively on fly ashes, while on soots, cement dust and volcanic ash less conversion of SO_2 is observed. Hoffmann (1983) investigated the role of light in aqueous phase oxidation catalyzed by hematite (α-Fe_2O_3). Under irradiation the presence of hematite enhances the oxidation rate through O_2 with more than an order of magnitude. Berresheim and Jaeschke (1983) investigated the conversion rate of SO_2 in the presence of several metal-containing particles. They found a significant anion influence. $MnCl_2$ is found to be most effective with a removal rate of ~3% h^{-1}. The chemical reactivity strongly depends on the amount of water condensed on the particles (Dlugi, 1983b; Dlugi et al., 1982). The conversion rate influenced by fly ashes and by high humidity in the near stack part of plumes can be high (tens of % h^{-1}). Mixing with plumes from cooling towers can enhance the distance where heterogeneous reactions occur up to a few kilometres (Nester et al., 1984). Because of the ion interactions in concentrated solutions the conversion rate of SO_2 in deliquescent salts can enhance several orders of magnitude compared to water droplets (Clarke, 1981; Clarke et al., 1983a; Clarke et al., 1983b). The amount of SO_2 oxidized, however, is small because of restrictions through the amount of water on the particles. Under ambient conditions the heterogeneous pathway for oxidation of SO_2 is considered to be of minor importance.

The relative importance of the various processes for the oxidation of SO_2 was evaluated theoretically by Middleton et al. (1980) and Möller (1980). These analyses lead to rough estimates only.

III.c.2. Field Data. Early measurements (before ~1973) of high SO_2 conversion rates in plumes are considered generally as unreliable (e.g. Wilson, 1978; Newman, 1981). Since then, an extended number of measurements of SO_2 conversion rates in coal or oil-fired power plant plumes and smelter plumes in the US, Canada and Australia are reported or reviewed (e.g. Whitby and Cantrell, 1978; Cantrell and Whitby, 1978; Husar et al., 1978; Meagher et al., 1978; Lusis et al., 1978; Hegg and Hobbs, 1980; Newman, 1981; Eatough et al., 1981; Williams et al., 1981; Meagher et al., 1981; Forrest et al., 1981; Garber et al., 1981; Gillani et al., 1981; McMurry et al., 1981;

Wilson and McMurry, 1981; Wilson, 1981; Zak, 1981). The conversion is investigated on diluted plumes from point sources, during transport up to distances of mostly a few tens of kilometres, sometimes more, up to a thousand kilometres (Williams et al., 1981). Most of the measurements are carried out under conditions favouring homogeneous gas phase conversion of SO_2 (dry or moderate dry (RH < 75%), no clouds). However, heterogeneous pathways are not excluded. Geographically most measurements are carried out at mid latitudes 20-45°. Observations at higher latitudes, however, agree (Lusis et al., 1978 (57°N); Elshout et al., 1978a (52°N); Gotaas, 1982 (50-60°N)).

In spite of wide geographical, seasonal, background and source variations a distinct diurnal variation in the conversion rate is observed. Typical daytime maximum rates up to 5% h^{-1} and nighttime rates of < 1% h^{-1} are measured (Newman, 1981; Wilson, 1981). The conversion of SO_2 shows a significant linear correlation with the solar radiation dose (Wilson, 1981). The mixing ratio of the plume with back ground air is important. At the fringes of the plume enhanced conversion rates are observed (Zak, 1981). The long-term averaged conversion rate of SO_2 in plumes entering into and mixing with clean air is generally less than 1% h^{-1}, but with polluted urban air the rate can be at least twice as fast (Newman, 1981). Plume-cloud interactions can result in high conversion rates (~10% h^{-1}) (Gillani et al., 1981; Newman, 1981).

Not only isolated plumes of single point sources have been investigated. The same conversion rates (0-5% h^{-1} gas phase, ~12% h^{-1} liquid phase) are observed in an urban plume (McMurry and Wilson, 1983) and in air parcels transported across England (~2% h^{-1} without clouds 5-20% h^{-1} with clouds) (Bamber et al., 1984).

III.c.3 Model Parameter Estimates. The gas phase conversion rate of SO_2 derived from an extended number of field measurements shows a diurnal variation with a low conversion rate at night and rates up to ~5% h^{-1} at noon during summer. When interaction with liquid water phase (e.g. by clouds, fog) occurs reaction rates are generally enhanced. The observations are reasonably consistent with theory. Using the reaction constant for the reaction of OH with SO_2 ($K = 1.4*10^3$ ppm^{-1} min^{-1}) and the hydroxyl concentrations given in Table 7, the 12-hour averaged conversion rate of the reaction of SO_2 with OH radicals can be estimated. In daytime these rates vary from 1.3% h^{-1} in summer to 0.2% h^{-1} in winter and are negligible (<0.1% h^{-1}) at night. The average rate over the year for this mechanism would be 0.3% h^{-1}. Tentatively the contribution of the oxidation of SO_2 in water droplets and on the surface of particles could be estimated at the same order of magnitude.

In Van den Hout et al. (1985) and Committee on Atmospheric Transport and Chemical Transformation in Acid Precipitation (1983) for more then twenty models conversion rates for SO_2 are given. Some models distinguish conversion rates for summer and winter and for diurnal variations. Typical values for diurnal variation are $0.2\text{-}3\%h^{-1}$ and $0.1 - 1.5\%\ h^{-1}$ for summer and winter conditions, respectively. Most models use a long-term averaged conversion rate value. Although a few models use lower ($\sim 0.5\%\ h^{-1}$) and higher values ($\sim 2\%\ h^{-1}$) a value of $1\%\ h^{-1}$ is generally used. So, the overall value of $1\%\ h^{-1}$ year-averaged is in reasonable accordance with theory, field measurements and the parameters used in most model studies.

III.d. Conversion of NO_x (Nitrogen Oxides)

III.d.1. Contributions to NO_x as measured by chemoluminescence. The term "NO_x" needs a more precise definition. Usually the important nitrogen oxides NO and NO_2 are meant. However, in the atmosphere other nitrogen oxides such as N_2O, NO_3, N_2O_5 and nitrogen compounds such as HNO_3 and HNO_2, inorganic nitrates as well as organic nitrates as peroxyacyl nitrates and nitrites, are found. The measurement of NO on the basis of the chemoluminescence reaction with ozone can be extended to the measurement of "NO_x" by catalytic reduction of higher-order oxidation products of nitrogen. With the exception of N_2O, the above-mentioned compounds are all measured, more or less efficiently, by such a procedure.
Van der Wiel (1977) showed that the measurement of NO, NO_2, PAN (peroxyacetylnitrate) and organic nitrite is quantitative even if a low-temperature (110°C) $FeSO_4$ converter is used. The same applies probably to HNO_2. An organic nitrate was measured for 20% only. Although experiments with HNO_3 were not quite reliable, it appears that this substance is detected only partially (personal communication, Van der Wiel). Bowermaster and Shaw (1981) report no response of HNO_3 with a 60°C $FeSO_4$ converter. Many chemoluminescent NO_x analysers in use are equipped with a high-temperature converter; the measurement of HNO_3 by such systems is reported to have a high efficiency (Winer et al., 1974; Spicer, 1976; personal communication Hollander/Verhagen).

III.d.2. Theoretical Studies and Laboratory Measurements. The chemistry of NO_x is rather complicated in view of the great number of different components. In recent years much knowledge of kinetics and mechanism of chemical transformation of NO_x has been obtained. There remain, however, key gaps in the understanding of atmospheric chemistry of NO_x. Recent reviews are given by Cox (1982), Committee on Atmospheric Transport and Chemical Transformation in Acid Precipitation (1983), Logan (1983), Calvert and Stockwell (1983b) and Atkinson and Lloyd (1984).

As seen in chapter 3.II.b the deposition velocity of NO and NO_2 is small compared to that of oxidation products as nitric acid.

Transformation processes therefore play an important role in the removal of NO_x species. A number of the most relevant reactions is compiled in Table 8. The reaction rate constants are, unless otherwise indicated, as recommended in a critical evaluation by Atkinson and Lloyd (1984) and are calculated for 15°C.

Emissions of NO_x are primarily in the form of NO. The conversion of NO into NO_2 proceeds in the dark by reaction of NO with ozone. In sunlight, an equilibrium on a time scale of minutes exists between the photolysis of NO_2 and the formation of NO_2 (reactions 1-3). At day time, the net conversion of NO to NO_2 is determined by the reaction of NO with peroxy radicals (HO_2, RO_2, RCO_3, see reactions 12, 17, 18). These radicals are formed during the degradation of hydrocarbons and aldehydes reacting with hydroxyl radicals and ozone. The involvement of several reactants and the occurrence of equilibria make the description of this chemistry in terms of first-order reactions highly inadequate. Modelling of NO/NO_2 with simple dispersion models like the Gaussian plume model is therefore not possible.

The concentration of the compounds RO_2NO_2 and HO_2NO_2 is determined by temperature-dependent equilibria (reactions 15, 16, 21, 22). They are considered to be only temporary sinks for NO_2 and radicals because they dissociate readily. Measurements of these compounds in ambient air have, to our knowledge, not been reported. The concentrations are believed to be rather low.

PAN originates from the reaction of organic peroxyacyl radicals (RCO_3) with NO_2. Because of its higher thermal stability it is a common photochemical product. The equilibrium determined by reactions 19 and 20 is not only strongly temperature-dependent, but depends also on the NO/NO_2 ratio (reaction 18 vs. 19). Therefore, the formation of PAN cannot be described by simple dispersion models either. Removal of PAN, however, is slow and especially under cold conditions PAN is transported over long distance.
With respect to nitrous acid, in the day time fast photolysis (reaction 11) combined with reaction 14 will lead to a low equilibrium concentration. It can be estimated to be:

$$[HNO_2] \cong [NO].[OH] \cdot \frac{k_{14}}{k_{11}}$$

For [OH] = 10^{-7} ppm and a light intensity of 100 Wm^{-2} this amounts to about 4% of the NO concentration. Measurements by Platt et al. (1980a), Platt et al. (1980b) indeed indicate that the concentration is very low (< 0.3 ppb) at day time. During the night, the concentration may increase reaching a maximum concentration of a few ppbv's just prior to sunrise. The production processes for this nocturnal formation of nitrous acid are not known. Heikes and

Table 8. Some reactions relevant to NO_x chemistry.

Reaction		k (ppm^{-1} min^{-1}) 288 K
1. $NO_2 + h\nu \rightarrow NO + O$		4.4 E-1 *
2. $O + O_2 + (M) \rightarrow O_3 + (M)$		2.4 E 1
3. $O_3 + NO \rightarrow NO_2 + O_2$		2.3 E 1
4. $O_3 + NO_2 \rightarrow NO_3 + O_2$		3.6 E-2
5. $NO_3 + NO \rightarrow 2\ NO_2$		2.8 E 4
6. $NO_3 + NO_2 \rightarrow N_2O_5$		1.7 E 3
7. $N_2O_5 \rightarrow NO_3 + NO_2$		9.9 E-1
8. $N_2O_5 + H_2O \rightarrow 2HNO_3$	heterogeneous	≦ 2 E-6
9. $NO + NO_2 + H_2O \rightarrow 2\ HNO_2$	"	≦ 1.6 E-11
10. $2\ HNO_2 \rightarrow NO + NO_2 + H_2O$	"	≦ 1.5 E-5
11. $HNO_2 + h\nu \rightarrow NO + OH$		2.6 E-2 *
12. $HO_2 + NO \rightarrow NO_2 + OH$		1.3 E 4
13. $OH + NO_2 \rightarrow HNO_3$		1.4 E 4
14. $OH + NO \rightarrow HNO_2$		1.0 E 4
15. $HO_2 + NO_2 \rightarrow HO_2NO_2$		1.9 E 3
16. $HO_2NO_2 \rightarrow HO_2 + NO_2$		1.5 E 0
17. $RO_2 + NO \rightarrow RO + NO_2$		1.2 E 4
18. $RCO_3 + NO \rightarrow RO_2 + NO_2$		1.2 E 4
19. $RCO_3 + NO_2 \rightarrow PAN$		8 E 3
20. $PAN \rightarrow RCO_3 + NO_2$		4.4 E-3
21. $RO_2 + NO_2 \rightarrow RO_2NO_2$		2.3 E 3
22. $RO_2NO_2 \rightarrow NO_2 + RO_2$		~2 E 1
23. $NO_3 + h\nu \rightarrow NO + O_2$		2.5 E 0 *
24. $NO_3 + h\nu \rightarrow NO_2 + O$		7.0 E 0 *
25. $2NO_2 + H_2O \rightarrow HNO_3 + HNO_2$		3 E-10
26. $2NO_{2(H_2O)} \rightarrow 2H^+ + NO_2^- + NO_3^-$	(liquid phase, Lee and Schwartz (1981))	
27. $NO_{(H_2O)} + NO_{2(H_2O)} \rightarrow 2H^+ + 2NO_2^-$	" " " " " (1981)	

* noon, summer, 50°N (~ 100 $W.m^{-2}$), Cox (1982)

Thompson (1983) investigated theoretically the potential role of wet particles in the HNO_2 chemistry. Homogeneous reactions (e.g. 14, 25) and heterogeneous and aqueous phase reactions (e.g. 9, 10, 26, 27) are one order of magnitude or more too slow to explain the night time production of nitrous acid. Other NO_x organic sources are proposed. The nitrous acid concentrations drop rapidly at sunrise because of the fast photolysis by reaction 11, to form NO and OH radicals.

It is generally assumed that nitric acid formation followed by wet or dry deposition is the dominant removal pathway of NO_x species. There are a number of possible reaction pathways which lead to nitric acid formation. Because of the uncertainties in the detailed mechanisms and the individual rate constants the apportionment of the several processes is not always clear. Analogous to SO_2, during conditions favourable for photochemical chemistry, the conversion of NO_2 with OH radicals is the dominant pathway (reaction 13). The reaction of OH radicals with NO_2 is about ten times as fast as the reaction with SO_2. Even in clouds during day time this pathway may account for the major part of the NO_2 conversion (Heikes and Thompson, 1983; Heikes and Thompson, 1984; Seigneur and Saxena, 1984). During night time, provided that other reaction pathways are inhibited, the formation of OH radicals e.g. by PAN decomposition and O_3 olefin reactions can also play a major role in the NO_2 conversion (Seigneur and Saxena, 1984). The reaction constant of reaction 13 (k_{13} = 1.4 E4 ppm^{-1} min^{-1}) and OH concentrations as given in Table 7 correspond with a day time NO_2 conversion rate of 13% h^{-1} in summer, ~2% h^{-1} in winter and ~3% h^{-1} in autumn and spring. The estimated conversion rate of NO_2 averaged over the year due to the reaction with OH radicals is 2.5% h^{-1}. The maximum conversion rate at noon on a sunny summer day may be over 30% h^{-1}.

The reaction of NO_2 with water vapour (reaction 25) is far too slow to play any significant role. Lee and Swartz (1981) investigated the aqueous reaction pathway of NO_2 (reactions 26 and 27). The presently available evidence indicates that because of the slow solubilities of NO and NO_2 at normal atmospheric conditions this pathway cannot be important (Committee on Atmospheric Transport and Chemical Transformation in Acid Precipitation, 1983; Cox, 1982). Another pathway of the conversion of NO_2 and the formation of HNO_3 is initiated by the reaction of NO_2 with O_3 leading to the formation of NO_3 radicals (reaction 4). The NO_3 radicals are in equilibrium with N_2O_5 through the fast reactions 6 and 7. During daylight hours NO_3 is rapidly photolyzed (reaction 23, 24) and when NO is present a fast destruction of NO_3 takes place (reaction 5) thus leading to low levels of NO_3 and N_2O_5. At night time or in clouds levels of NO_3 and N_2O_5 may rise and the heterogeneous reaction of N_2O_5 on wet surfaces leads to nitric acid formation (reaction 8). There are much uncertainties with regard to this reaction pathway. Though there is no absolute evidence, mostly it is assumed that the reaction of N_2O_5 with water is heterogeneous and the upper limit of the reaction seems to be 2 * 10^{-6} ppm^{-1} min^{-1} (Atkinson and Lloyd, 1984).

Direct measurement of NO_3 concentrations confirms this theory (Platt and Perner, 1980b; Platt et al., 1980c; Platt et al., 1984). The levels of NO_3 observed at night indicate the appearance of strong nocturnal loss processes. The reactions with olefins, even when the upper limit of reactivity is used, are not sufficient to account for the loss processes (Heikes and Thompson, 1983). Because of the rapid decrease of the lifetime of NO_3 with increasing relative humidity the involvement of water in the loss processes seems plausible (Platt et al., 1984). Richards (1983) proposed that the heterogeneous loss of N_2O_5 could be the major nocturnal loss process. Modelling Richard's mechanism predicts the correct order of magnitude of observed night time NO_3 concentrations (Jones and Seinfeld, 1983). The introduction of this heterogeneous pathway results in a substantial decrease of the nocturnal NO_3 and N_2O_5 levels and accounts for a considerable fraction of the nitrate produced (Sverdrup and Hov, 1984).

Several investigators modelled the formation of nitric acid under different atmospheric conditions. When photochemical activity is high the conversion of NO_2 with OH radicals is the major pathway. However, under conditions with low photochemical activity and the presence of wet surfaces, for example in dense fog, in optically thick clouds and winter clouds and at night, the pathway that leads to NO_3 and N_2O_5 formation and the heterogeneous reactions of N_2O_5 and/or NO_3 are important and can be the dominant conversion process of NO_2 (Jones and Seinfeld, 1983; Heikes and Thompson, 1983; Seigneur and Saxena, 1984; Sverdrup and Hov, 1984). Calvert and Stockwell (1983b) modelled the diurnal variation of the transformation processes in polluted air. They treated the H_2O-N_2O_5 reaction as a homogeneous one and varied the reaction rate from zero up to 2*10^{-5} ppm^{-1} min^{-1}. At the highest value of the rate constant the night time conversion of NO_2 by this reaction is more important than the day time conversion by OH radicals. For the night time conversion rate values of 10% h^{-1} for moderately and up to 80% h^{-1} for heavily polluted atmospheres are found. These values result in 24-hour averaged values of ~ 20% h^{-1}. Even at a reaction rate of zero, night time conversion of NO_2 to HNO_3 was considerable (a few % h^{-1}) due to reactions of the formed NO_3 radicals with aldehydes. In this case the 24-hour averaged NO_2 conversion is 5-10% h^{-1}. Calvert and Stockwell suggest that the nocturnal conversion of NO_2 takes place through the aldehyde reactions rather than through the N_2O_5-H_2O reaction.

Despite the many uncertainties about the mechanisms, the potential of the NO_2 conversion to nitric acid through the NO_3 pathway has to be taken into account. If we assume that the reactions forming nitric acid via NO_3 are fast relative to the formation of NO_3 an approximation can be made for this pathway from the rate constant of reaction 4. A concentration of 10 ppb or 50 ppb of O_3 results in a conversion rate of NO_2 of 4% h^{-1} and 22% h^{-1}, respectively. Tentatively the contribution of this pathway could be estimated at the same order of magnitude as the conversion of NO_2 by OH radicals.

All conversion rates are given relative to NO_2. If NO is present, the conversion rates relative to NO_x are lower. Furthermore, the presence of NO inhibits the conversion pathways via NO_3 because of the fast destruction of NO_3 through the reaction with NO. The variability of the possible conditions and the great influence of these conditions on the various pathways make an estimation of an overall averaged conversion rate of NO_x, including the different conversion pathways, too speculative.

III.d.3. Field Data. Few authors have tried to estimate conversion rates for NO_x from measuring data. Chang et al. (1979) estimated from measurements of the NO_x/CO-ratio in the Los Angeles Basin that the year-averaged removal rate of NO_x at least amounts to 4% per hour during day time. These authors deduce similar rates from NO_x, PAN and HNO_3 measurement during summer by Spicer in St. Louis and in Los Angeles. Guicherit et al. (1980) estimated from the NO_x/CO ratio in Delft, The Netherlands, day time NO_x removal rates of 1% h^{-1} in winter, 5% h^{-1} in summer and 14% h^{-1} during photochemical episodes. Spicer (1980, 1982) deduced from measurements of NO_x and of tracer gases that the conversion rate in the urban plume of the city of Boston amounted to 14 - 24% h^{-1} during dispersion over the ocean. For similar measurements in the plume of Phoenix, Arizona, under very hot and dry conditions he found an NO_x upper-limit removal rate of 5% h^{-1}. An explanation for this discrepancy is lacking. Forrest et al. (1981) measured the conversion rate of NO_x in the Cumberland power plant plume. He found conversion rates of 0.1 - 6% h^{-1} of NO_x during night time and early morning and values of 0.6 - 22% h^{-1} of NO_x during late morning and afternoon. A dominant role of the OH radical in day time conversion is assumed, because of the constant ratio of the rate of oxidation of NO_x and SO_2. Measurement by Hegg et al. (1984a) of the formation of nitrate in cloud water are consistent with simulations with the numerical model of Heikes and Thompson (1983), suggesting formation of nitric acid through the NO_3-N_2O_5 pathway. Measurements by Lazrus et al. (1983) of the generation of HNO_3 in clouds in the Ohio River Valley during autumn has been simulated by several investigators. Heikes and Thompson (1983) found that the nitric acid formation is the result of both the OH pathway and the heterogeneous NO_3-N_2O_5 pathway, Seigneur and Saxena (1984) concluded that the formation of HNO_3 under the same conditions in clouds predominantly is the result of the NO_3-N_2O_5 reaction pathway.

Many measurements of the ambient concentrations of oxidation products of NO_x have been reported. High ratios of the concentration of the oxidation products versus the NO_x concentration indicate high conversion rates of NO_x (Spicer, 1977; Grosjean, 1983). Accurate values of the conversion rate of NO_x cannot be deduced from these measurements.

III.d.4. Model Parameter Estimates. Summarizing, it appears that a year-averaged conversion rate of 2.5% h^{-1} for NO_2 due to the reaction with OH radicals is a reasonable supposition. In modelling studies, if possible, the seasonal variation and the day-night variation resulting from the estimated OH concentration according to Table 7, have to be taken into account. If NO is present the conversion rates relative to NO_x are consequently lower. Despite the many uncertainties about reaction mechanisms and rate constants, the potential of the NO_2 conversion to nitric acid through the NO_3 pathway has to be taken into account. Especially during night time, in clouds or fog and during winter this pathway may lead to considerable amounts of nitric acid. Tentatively the contribution of this pathway could be estimated to be of the same order of magnitude as the conversion of NO_2 by OH radicals. However, in the presence of NO or solar radiation the conversion pathway via NO_3 is inhibited because of the fast destruction of NO_3 by the reaction with NO and the fast photolysis of NO_3.

IV. REMOVAL OF POLLUTANTS IN PLUMES

IV.a. Introduction

There are several reasons to consider the removal of pollutants in plumes as a separate case. Plumes can be defined in a broad sense as air parcels in ambient air that can still be recognized as originating from a certain source or source complex. In a more narrow sense, plumes can be considered to originate from large point sources, often high stacks. Plumes therefore differ from their surroundings in the first place by the higher concentration of pollutants. This implies that the concentration of reactants is primarily determined by the plume itself; chemical conversion in such a situation could be significantly different from conversion in a situation where the reactant concentration is determined by some average value for an area.

A well-known example is the conversion of NO by ozone in power plant plumes (Elshout, 1978b). Since the ozone in the plume is depleted by rapid reaction with the high concentration of NO, the conversion rate is determined by diffusion of ambient air into the plume. Another aspect to be considered is dry deposition. From plume studies in the USA (Wilson, 1978) it is found that emissions from high stacks are often transported during the night above an inversion, and therefore are decoupled from the ground. Dry deposition is negligible in such cases (Husar, 1978). Transport of SO_2 above the mixing layer has been shown to contribute heavily to the flux of this pollutant into the Netherlands, indicating the importance of this mechanism for emissions from tall stacks (V. Egmond et al., 1978).

Apart from this aspect, it has to be recognized that taking into account the surface removal by dry deposition in modelling studies by reducing the source strength as a function of distance ("source depletion model") may lead to serious errors in calculated ground concentrations. This has become clear from a comparison with so-called "surface depletion models" that account for non-instantaneous dispersion of the concentration loss across the plume (Horst, 1977; Draxler and Elliott (1977); Berkowics and Prahm (1978); Horst et al., 1983b). In order to establish possible plume-specific parameter values a literature investigation was performed for SO_2 and NO_x.

IV.b. Sulphur Dioxide

In older plume studies, enhanced conversion rates for SO_2 are reported (Levy et al., 1976). These studies, however, are considered to be of limited value owing to less reliable measuring techniques. According to Wilson et al. (1976), who reviewed EPA plume studies, conversion in the first kilometres of plumes is rather low. Due to the slow rate, Smith et al. (1975) were not able to measure the conversion in a power plant plume. Lusis et al. (1978) found a conversion of SO_2 < 0.5% h^{-1} in the absence of sunlight. In sunshine the conversion increased to 1 - 3% h^{-1}. Meagher et al. (1978) found an average conversion of SO_2 to sulphate of 1.1% per hour with a maximum of 4.3%. Oxidation mostly occurred quite near the stack. The rapid decrease of oxidation rate was attributed by the authors to the fast decrease of available surface of particulates or to the poisoning of catalyst present on these particulates. The average oxidation rate at a distance of more than 10 km was 0.2% h^{-1}. Also by Lusis and Wiebe (1976) a slight decrease of the oxidation rate was observed in a plume of a nickel smelter. They found oxidation rates generally less than 3% h^{-1} with an averaged value of 1% h^{-1}. Schwartz and Newman (1978) also assume that the rapid decrease of the SO_2 oxidation rate found in some of their studies is caused by the decrease of catalyst particles or as a result of plume expansion or by catalyst poisoning. Forrest and Newman (1977a) found that the oxidation of SO_2 seldom exceeded 5% even for distances up to 70 km and for times of transport of 200 minutes. They could not ascertain a clear correlation of the SO_2 oxidation with distance, time of transport, temperature, relative humidity, time of the day or stability of the atmosphere. Elshout et al. (1979) found a conversion of SO_2 of about 1% h^{-1}. Husar et al. (1978) determined the summer sulphur budget of a power plant plume near St. Louis (USA). They found a conversion of SO_2 between 1 and 4% per hour about noon and < 0.5% h^{-1} during the night. Forrest and Newman (1977a, 1977b) found corresponding low values. Wilson et al. (1976), studying chemical reactions in power plant plumes, observed that later in the plume a sharp increase occurred in the rate of conversion of SO_2 to sulphate. The model calculations of Miller et al. (1978) also predict an increased production of sulphate aerosol in an aged plume. Calvert et al. (1978) theoretically found that under conditions of highly NO_2-polluted atmosphere, such as present in the early stages of dilution of a stack gas plume, a significant conversion (up to 1.4% h^{-1}) of SO_2 by radicals, e.g. the O(3P)-atom, may occur.

In the literature it is suggested that there is an important difference in the conversion rate of SO_2 in oil-fired and in coal-fired power plants. This idea originates from studies of Newman et al., who found high sulphate levels in the plume of an oil-fired power plant (Newman et al., 1975). The isotope technique used by these authors has, however, been characterized as erroneous by Wilson, while the data analysis, from which it is concluded that second-order processes play a role, has been in dispute (Overton, 1978). However, it is interesting to note that Honolya and Fortune (1978) show that the content of water-soluble sulphate is 20 times as high in fly ash from oil-fired power plants as in fly ash from coal-fired power plants. This could mean that sulphate is formed already inside the stack. In a more recent review of plume studies, Newman confirms the conception that an oil-fired plume generally may not be more reactive than a coal-fired plume. However, primary sulphate emissions of oil-fired plants can be higher, which is caused by power plant operating conditions, and consequently a higher ambient reactivity could be possible (Newman, 1981). Primary sulphate emissions are observed in smelter plumes as well as in coal-fired and oil-fired power plant plumes (Eatough et al., 1981; Forrest et al., 1981; Garber et al., 1981). The amounts of primary sulphate vary, typical values up to a few percent of the total sulphur are reported. Generally during the first hour of transport the sulphate in the plume is dominated by the primary sulphate (Eatough, 1981). In the review of plume studies of Newman (1981), the possibility of an initial induction period with little oxidation as well as enhanced oxidation rates in the vicinity of the stack are mentioned. It may be concluded that firm evidence for enhanced chemical conversion in plumes is lacking.

Removal of SO_2 in a plume may be due to deposition as well. A very large decrease was reported by H. Flyger et al. (1977) and by H. Flyger (1978). The overall half-life of SO_2 in the plume of an oil-fired power plant appeared to be only half an hour. Bingemer

(1976) found in the "plume" of the Ruhr district a removal of SO_2 of 10 - 25% per hour. Possibly, deposition in densely built areas such as the Ruhr district is enhanced. Georgii and Perseke (1979) found in non-polluted areas about equal values for dry and wet deposition of SO_2, but noticed that dry deposition in the Ruhr district was increased by a factor of 3 - 7, whilst wet deposition was about equal. Husar et al. (1978) noticed that deposition losses from the plume of a large power plant during the day amounted to 10% per hour at most. These losses were negligible during the night. On the basis of these limited amounts of data, no firm conclusions on deposition losses can be drawn.

IV.c. Nitrogen Oxides

The nitrogen oxides emitted by large power plants consist for 95 - 98% of NO. In the immediate neighbourhood of the stack reaction of NO with oxygen to NO_2 is dominant. The rate of this reaction, which is second order in NO, rapidly decreases owing to dilution. Elshout (1978b) estimates from model calculations that for power plant plumes in the Netherlands, the total conversion by this mechanism amounts to a few percent, and takes place within a few kilometres from the stack. At larger travel distances, oxidation of NO by ozone in the ambient air is dominant. The reaction rate is determined by diffusion, so that oxidation takes place rather slowly. The conversion occurs especially at the edges of the plume where mixing with the surrounding air takes place (Meagher et al., 1978). A number of investigators have carried out model calculations of this process (Kewley et al., 1978; White, 1977; Lusis, 1978; Melo et al., 1978; Williams, 1976; O'Brien et al., 1976a; O'Brient et al., 1976b; Carmichael and Peters, 1981; Shu et al., 1978; Builtjes, 1981; Persson, 1983). Both theory and measurements show that in case the surrounding air is not polluted, the NO_2/NO ratio does not increase strongly. Elshout (1978b) measured values not much higher than 2. The highest value measured by Hegg et al. (1977) was 4.3. During the oxidation of NO to NO_2 ozone is consumed. However, it has been suggested that in a plume ozone can also be formed. Literature data on this topic are contradictory. Davis (1974) found an ozone bulge at greater distance from a stack. He attributed this to the occurrence of radical reactions in the plume. Further analyses seem to indicate that the ozone bulge must have been caused by mixing with air with reactive hydrocarbons (See Miller et al., 1978). However, the formation of ozone in plumes may occur under certain circumstances. Elshout (1979) observed the formation of ozone during strong solar radiation (> 38 $J.cm^{-2}.h^{-1}$).

Since chemical conversion of NO_x is determined by reaction of NO_2, and NO_2 dry deposition is much more efficient than NO dry deposition, it is expected that chemical conversion and deposition of NO_x in plumes is generally lower than in ambient air owing to the relatively high NO/NO_2 ratio.

V. LITERATURE

Aalst, R.M. van, 1982, Dry deposition of NO_x. In: "Air pollution by nitrogen oxides", (T. Schneider and L. Grant, eds.) p. 263. Elsevier, Amsterdam.

Aalst, R.M. van, Builtjes, P.J.H., Heidema, L.F., Meijer, G.M., 1983a, Parametrising dry deposition of gases - laboratory techniques and field experiments. Presented on NATO-CCMS 14th I.T.M. on air pollution modelling and its application. Copenhagen, September 1983.

Aalst, R.M. van, Diederen, H.S.M.A., Duyzer, J.H. and Meijer, G.M., 1983b, Dry deposition of acid precursors in the Netherlands. VDI-berichte 500, p. 97.

Adamowicz, R.F., 1979, A model for the reversible washout of sulphur dioxide, ammonia and carbon dioxide from a pollutial atmosphere and the production of sulphates in raindrops. Atmospheric Environment 13, 105.

Atkinson, R. and Lloyd, A.C., 1984, Evaluation of kinetic and mechanistic data for modelling of photochemical smog. J. Phys. Chem. Ref. Data 13, 315.

Bache, D.H., 1979a, Particle transport within plant canopies. I. A frame work for analysis. Atmospheric Environment 13, 1257.

Bache, D.H., 1979b, Particulate transport within plant canopies. II. Prediction of deposition velocities. Atmospheric Environment 13, 1681.

Bache, D.H., 1984, Prediction of the bulk deposition velocity and concentration profiles within plant canopies. Atmospheric Environment 18, 2517.

Bamber, D.J., Clark, P.A., Glover, G.M., Healey, P.G.W., Kallend, A.S., Marsh, A.R.W., Tuck, A.F. and Vaughan, G., 1984, Air sampling flights round the British isles at low altitudes: SO_2 oxidation and removal rates. Atmospheric Environment 18, 1777.

Barrie, L.A., 1978, An improved model of reversible SO_2 washout by rain. Atmospheric Environment 12, 407.

Barrie, L.A., 1981, The prediction of rain acidity and SO_2 scavenging in eastern North America. Atmospheric Environment 15, 31.

Barrie, L.A., and Neustadter, J., 1983, The dependence of sulphate scavenging ratios on meteorological variables. In: "Precipitation scavenging, dry deposition and resuspension" (Pruppacher et al., (eds.)), Elsevier Science Publ., New York, 1983.

Barrie, L.A., and Walmsley, J.L., 1978, A study of sulphur dioxide deposition velocities to snow in Northern Canada. Atmospheric Environment, 12, 2321.

Beilke, S., and Gravenhorst, G., 1978, Heterogeneous SO_2 oxidation in the droplet phase. Atmospheric Environment 12, 231.

Beilke, S., 1970, Laboratory investigations of washout of trace gases. Proc. Symp. on Precipitation Scavenging USAEC Symp. Services No. 22, p. 261, 1970.

Bengtson, C. et al., 1981, Deposition and uptake of nitrogen oxides in scots pine needles. Institutet för Vatten- och Luftvårds-forskning, report IVL B647, Uppsala, Sweden.

Berkowicz, R., and Prahm, L.P., 1978, Pseudospectral simulation of dry deposition from a point source, Atmospheric Environment, 12, 379.

Berresheim, H., and Jaeschke, W., 1983, Laboratory Studies of the chemical removal of SO_2 in the presence of metal containing particles. Paper presented at the Advanced Study Institute on Chemistry of Multiphase Atmospheric Systems, Corfu, 1983.

Bingemer, H., et al., 1976, Investigations of medium range transport of SO_2 and sulphate by aeroplane measurements. Paper presented at Risö, Danmark, June 1976. Project COST 61 a.

Bos, R., Goudena, E.J.G., Guicherit, R., Hoogeveen, A., and De Vreede, J.A.F., 1978, Atmospheric precursors and oxidants concentrations in the Netherlands, In: "Photochemical smog formation in the Netherlands" (R. Guicherit ed.), p. 20, TNO, 's-Gravenhage.

Böttger, A. et al., 1980, Atmosphärische Kreisläufe von Stickstoff-oxiden und Ammoniak. Bericht der Kernforschungsanlage Jülich, Jül-1558, Nov. 1978.

Bowermaster, J., and Shaw jr., R.W., 1981, A source of gaseous HNO_3 and its transmission efficiency through various materials, Journ. Air. Poll. Contr. Asc., 31, 788.

Breemen, N. van, Burrough, B.A., Velthorst, E.J., Dobben, H.F. van, Wit, T. de, Ridder, T.B. and Reijnders, H.F.R., 1982, Soil acidification from atmospheric ammonium sulphate in forest canopy throughfall. Nature 299, 548.

Brunner, G., Vreede, J.A.F. and Guicherit, R., 1978, UV measurements as a parameter of photochemical smog formation. In: "Photochemical smog formation in the Netherlands", (R. Guicherit ed.), p. 68, TNO, 's-Gravenhage.

Builtjes, P.J.H., 1981, A comparison between chemically reacting plumes and wind tunnel experiments. 12th I.T.M. on Air Pollution Modelling and its Applications. Palo Alto, USA (1981).

Calvert, J.G., Su, F., Bottenheim, J.W. and Strausz, O.P., 1978, Mechanism of the homogeneous oxidation of sulphur dioxide in the troposphere. Atmospheric Environment 12, 197.

Calvert, J.G., and Stockwell, W.R., 1983a, In: "Acid precipitation: SO_2, NO and NO_2 oxidation mechanism: Atmospheric considerations". Ann Arbor Scientific Publications, in press.

Calvert, J.G., and Stockwell, W.R., 1983b, Acid generation in the troposphere by gas-phase chemistry. Environmental Science and Technology, 17, 428A.

Campbell, M.J., Sheppard, J.C., and Au, B.F., 1979, Measurement of hydroxyl concentration in boundary layer air by monitoring CO oxidation. Geophys. Res. Lett., 6, 175.

Campbell, M.J., Sheppard, J.C., Hopper, F.J. and Hardy, R., 1982, Measurements of tropospheric hydroxyl radical concentrations by ^{14}C tracer method, 2nd Symposium on the composition of the Nonurban Troposphere, Williamsburg, Va., 1982.

Cantrell, B.K., and Whitby, K.T., 1978, Aerosol size distribution and aerosol volume formation for a coal-fired power plant plume, Atmospheric Environment, 12, 323.

Carmichael, G.R. and Peters, L.K., 1981, Application of the mixing-reaction in series model to NO_x-O_3 plume chemistry. Atmospheric Environment, 15, 1069.

Carmichael, G.R. and Reda, M., 1982, Non-isothermal SO_2 absorption by water droplets - The effects of precipitation intensity, sulphate aerosol scavenging and aqueous S(IV) oxidation. Atmospheric Environment 16, 2905.

Carter, W.P.L., Atkinson, R., Winer, A.M., and Pitts jr., J.N., 1982, Experimental investigation of chamber dependent radical sources, Int. J. Chem. Kinet. 14, 1071.

Chamberlain, A.C., 1967, Transport of Lycopodium spores and other small particles to rough surfaces. Proc. Royal Soc. London, 296, 45.

Chamberlain, A.C., 1983, Deposition and resuspension. In: "Precipitation scavenging, Dry Deposition and Resuspension" (H.R. Pruppacher et al., eds.) Vol. 2, p. 731, Elsevier, New York, 1983.

Chang, T.Y., Norbeck, J.M., and Weinstock, B., 1979, An estimate of the NO_x removal rate in an urban atmosphere. Environmental Science and Technology, 13, 1534.

Charlson, R.J. and Rodhe, H., 1982, Factors controlling the acidity of natural rainwater. Nature, 295, 683.

Clarke, A.G., 1981, Electrolyte solution theory and the oxidation rate of sulphur dioxide in water. Atmospheric Environment, 15, 1591.

Clarke, A.G., and Williams, P.T., 1983a, The oxidation of sulphur dioxide in electrolyte droplets. Atmospheric Environment, 17, 607.

Clarke, A.G. and Radojevic, M., 1983b, Chloride ion effects on the aqueous oxidation of SO_2. Atmospheric Environment, 17, 617.

Clough, W.S., 1975, The deposition of particles on moss and grass surfaces. Atmospheric Environment 9, 1113.

Committee on Atmospheric Transport and Chemical Transformation in Acid Precipitation, 1983, Acid Deposition, Atmospheric Processes in Eastern North America. National Academy Press, Washington, 1983.

Cox, R.A., 1982, Chemical transformation processes for NO_x species in the atmosphere; In: "Air Pollution by Nitrogen Oxides" (T. Schneider and L. Grant, eds.), Elsevier Scientific Publishing Company, Amsterdam, 1982.

Dana, M.T., and Hales, J.M., 1976, Statistical aspects of the washout of polydisperse aerosols. Atmospheric Environment 10, 45.

Dasch, J.M., 1983, A comparison of surrogate surfaces for dry deposition collection. In: "Precipitation Scavenging, Dry Deposition and Resuspension" (H.R. Pruppacher et al., eds.) Vol. 2, p. 883, Elsevier, New York, 1983.

Daum, P.H., Schwartz, S.E. and Newman, L., 1983, Studies of the gas and aqueous phase composition of stratiform clouds. In: "Precipitation Scavenging, Dry Deposition, and Resuspension" (Pruppacher et al. (eds.)). Elsevier Science Publishing Co., New York, 1983.

Davenport, H.M. and Peters, L.K., 1978, Field studies of atmospheric particulate concentration changes during precipitation. Atmospheric Environment 12, 997.

Davidson, C.I., Goold, W.D., and Wiersma, G.B., 1983, Dry deposition of trace elements in Olympic National Park in: "Precipitation Scavenging, Dry Deposition and Resuspension" (H.R. Pruppacher et al., eds.), Vol. 2, p, 871. Elsevier, New York, 1983.

Davidson, C.I., Lindberg, S.E., Schmidt, J.A., Cartwright, L.G., Landis, L.R., 1985, Dry deposition of sulphate onto surrogate surfaces. J. Geophys. Res. 90, D1, 2123.

Davidson, C.I., Miller, J.M. and Pleskov, M.A., 1982, The influence of surface structure on predicted particle dry deposition to natural grass canopies. Water, Air and Soil Pollution 18, 25.

Davies, T.D., 1976, Precipitation scavenging of sulphur dioxide in an industrial area. Atmospheric Environment 10, 879.

Davies, T.D., 1979, Dissolved sulphur dioxide and sulphate in urban and rural precipitation (Norfolk, U.K.). Atmospheric Environment 13, 1275.

Davies, T.D., 1983, Sulphur dioxide precipitation scavenging. Atmospheric Environment 17, 797.

Davies, T.D., and Mitchell, J.R., 1983, Dry deposition of sulphur dioxide onto grass in rural eastern England. In: "Precipitation Scavenging, Dry Deposition and Resuspension" (H.R. Pruppacher et al., eds.) Vol. 2, Elsevier, New York.

Davies, T.D., and Nicholson, 1982, Dry deposition velocities of aerosol sulphate in rural Eastern England. In: "Deposition of atmospheric pollutants" (H.W. Georgii and J. Pankrath, eds.), p. 31. D. Reidel, Dordrecht.

Davis, C.S., and Wright, R.G., 1985, Sulphur dioxide deposition velocity by a concentration gradient measurement system. J. Geophys. Res. 90, D1, 2091.

Davis, D.D., Smith, G., and Klauber, G., 1974, Trace gas analysis of power plant plumes via aircraft measurement: O_3, NO_x and SO_2 chemistry. Science, 186, 733.

Davis, D.D., Heaps, W.S., Philen, D., Rodgers, M.O., McMagee, T., Nelson, A., and Moriarty, A.J., 1979a, Airborne laser induced fluorescence system for measuring OH and other trace gases in the parts - per - quadrillion to parts - per - trillion range. Rev. Sci.Instrum., 50, 1505.

Davis, D.D., Heaps, W., Philen, D. and McGee, T., 1979b, Boundary layer measurements of the OH radical in the vicinity of an isolated power plant plume: SO_2 and NO_2 chemical conversion times. Atmospheric Environment, 13, 1197.

Davis, L.I., Wang, C.C., Tang, X., Niki, H., and Weinstock, B., 1982, Fluorescence measurements of OH at Niwot Ridge, 2nd Symposium on the Composition of the Nonurban Troposphere, Williamsburg, Va., 1982.

Delany, A.C., Davies, T.D., 1983, Dry deposition of NO_x to grass in rural east Anglia. Atmospheric Environment 17, 1391.

Derwent, R.G., et al., 1979, Computer modelling studies of photochemical air pollution formation in North West Europe. AERE Harwell Report R - 9434, HMSO, London.

Dlugi, R., 1983a, The oxidation of sulphur dioxide in water containing aerosol particles and droplets. Paper presented at the Advanced Study Institute on Chemistry of Multiphase Atmospheric Systems, Corfu, 1983.

Dlugi, R., 1983b, SO_2 oxidation in aerosol particles and droplets. J. Aerosol Sci. 14, 292.

Dlugi, R., Jordan, S., and Lindemann, E., 1982, Influence of particles properties on heterogeneous SO_2 reactions. In: "Physico-chemical behaviour of atmospheric pollutants" (B. Versino and H. Ott, eds.), p. 308, Reidel, Dordrecht (1982).

Doran, J.C. and Droppo, J.G., 1983, Profiles of elements in the surface boundary layer. In: "Precipitation Scavenging, Dry Deposition and Resuspension" (H.R. Pruppacher et al., eds.) Vol. 2, p. 1003, Elsevier, New York.

Draxler, R.R., and Elliott, W.P., 1977, Long range travel of air borne material subjected to dry deposition, Atmospheric Environment, 11, 35.

Duyzer, J.H., Meijer, G.M., Aalst, R.M. van, 1983, Measurement of dry deposition velocities of NO, NO_2 and O_3 and the influence of chemical reactions. Atmospheric Environment 17, 2117.

Eatough, D.J., Richter, B.E., Eatough, N.L., and Hansen, L.D., 1981, Sulphur chemistry in smelter and power plant plumes in the western U.S. Atmospheric Environment, 15, 2241.

Eggleton, A.E.J., and Cox, R.A., 1978, Homogeneous oxidation of sulphur compounds in the atmosphere, Atmospheric Environment, 12, 227.

Egmond, N.D. van, Tissing, O., Onderdelinden, D. and Bartels, C., 1978, Quantitative evaluation of mesoscale air pollution transport, Atmospheric Environment, 12, 2279.

Elshout, A.J., Viljeer, J.W. and Duuren, H. van, 1978a, Sulphates and sulphuric acid in the atmosphere in the years 1971 - 1976 in the Netherlands, Atmospheric Environment, 12, 785.

Elshout, A.J., et al., 1978b, De oxidatie van stikstofmonoxide in rookpluimen (The oxidation of nitrogen monoxide in plumes). Electrotechniek, 56, 429.

Elshout, A.J., Duuren, H. van, Römer, F.G. and Viljeer, J.W., 1979, Messungen aus Flugzeugen: Verbreitung und Umwandlung primärer luftfremder Komponente in Rauchfahnen. "European Symposium Physico-chemical behaviour of atmospheric pollutants", Ispra 1979. Proceedings, B. Versino and H. Ott, editors, p. 451.

Engelmann, R.Y., 1970, Scavenging prediction using ratios of concentrations in air and precipitation. Proc. Symposium on precipitation scavenging. A.E.C. Symposium Series 22, 475-485.

E.P.A., 1977, Uses, limitations and technical basis of procedures for quantifying relationships between photochemical oxidants and precursors. Report E.P.A. -450/2 - 77 - 021 a.

Everett, R.G., Hicks, B.B., Berg, W.W., and Winchester, J.W., 1979, An analysis of particulate sulphur and lead gradient data collected at Argonne National Laboratory. Atmospheric Environment, 13, 931.

Fisher, B.E.A., 1982, The transport and removal of sulphur dioxide in a rain system. Atmospheric Environment 16, 775.

Flothmann, D., Jähne, B., Klöpffer, W., Knacker, Th., Marggrander, E., Platt, U., Schaub, H. and D. Wagenbach, 1982a, Austausch von Luftverunreinigungen an der Grenzfläche Atmosphäre / Erdoberfläche (trockene Deposition). Zwischenbericht zum Teilprojekt 1: Deposition von Gasen. Battelle-Institut Frankfurt, report BleV-R-64.284-2.

Flothmann, D., Wagenbach, D., Rödel, W., Marggrander, E., and Bol, J., 1982b, Austausch von Luftverunreinigungen an der Grenzfläche Atmosphäre/Erdoberfläche (trockene Deposition). Zwischenbericht zum Teilprojekt 1: Deposition von Aerosolen. Battelle-Institut Frankfurt, report BleV-R-64.284-3.

Flyger, H., et al., 1977, Physical and chemical processes of sulphur dioxide in the plume of an oil-fired power station. Risø Rep. (Den. Res. Establ. Risø) 1977, 328.

Flyger, H., Lewin, E., Lund Thomsen, E., Fenger, J., Lyck, E. and Gryning, S.E., 1978, Airborne investigations of SO_2 oxidation in the plumes from power stations. Atmospheric Environment, 12, 295.

Forrest, J., Garber, R.W. and Newman, L., 1981, Conversion rates in power plant plumes based on filter pack data: the coal-fired Cumberland plume. Atmospheric Environment, 15, 2273.

Forrest, J., and Newman, L., 1977a, Further studies on the oxidation of sulphur dioxide in coal-fired power plant plumes. Atmospheric Environment, 11, 465.

Forrest, J., and Newman, L., 1977b, Oxidation of SO_2 in the Sudbury smelter plume. Atmospheric Environment, 11, 517.

Fowler, D., 1978, Dry deposition of SO_2 on agricultural crops. Atmosph. Env., 12, 369.

Fowler, D., 1980, Removal of sulphur and nitrogen compounds from the atmosphere in rain and by dry deposition, In: D. Drabløs, A. Tollan (eds.), "Ecological impact of acid precipitation", Oslo, 1980.

Fowler, D., 1984, Transfer to terrestrial surfaces. Phil. Trans. R. Soc. Lond. B305, 381.

Fowler, D., and Cape, N., 1983, Dry deposition of SO_2 onto a scots pine forest. In: "Precipitation Scavenging, Dry Deposition and Resuspension" (H.R. Pruppacher et al., eds.), Vol. 2, p. 763, Elsevier, New York.

Fowler, D., and Unsworth, M.H., 1979, Turbulent transfer of sulphur dioxide to a wheat crop. Quart. J. R. Met. Soc. 105, 767.

Freiberg, J.E., and Schwartz, S.E., 1981, Oxidation of SO_2 in aqueous droplets: mass transport limitation in laboratory studies and the ambient atmosphere. Atmospheric Environment 15, 1145.

Galbally, I.E., Garland, J.A. and Wilson, M.J.G., 1979, Sulphur uptake from the atmosphere by forest and farmland. Nature 280, 49.

Garber, R.W., Forrest, J. and Newman, L., 1981, Conversion rates in power plant plumes, based on filter pack data: the oil-fired Northport plume. Atmospheric Environment 15, 2283.

Garland, J.A., 1977, The dry deposition of sulphur dioxide to land and water surfaces. Proc. R. Soc. Lond. A 354, 245.

Garland, J.A., 1978, Dry and wet removal of sulphur from the atmosphere. Atmospheric Environment 12, 349.

Garland, J.A., Penkett, S.A., 1976, Absorption of peroxy acetyl nitrate and ozone by natural surfaces. Atmospheric Environment 10, 1127.

Garland, J.A., and Cox, L.C., 1982, Deposition of small particles to grass. Atmospheric Environment 16, 2699.

Garland, J.A., 1983, Principles of dry deposition: Application to acidic species and ozone. VDI-Berichte 500, p. 83.

Georgiï, H.W. and Perseke, C., 1979, Some results on wet and dry deposition of sulphur compounds. "European Symposium Physico-chemical behaviour of atmospheric pollutants", Ispra, 1979. Proceedings, B. Versino, H. Ott, editors, p. 410.

Gillani, N.V., Kohli, S. and Wilson, W.E., 1981, Gas-to-particle conversion of sulphur in power plant plumes - I. Parametrization of the conversion rate for dry, moderately polluted ambient conditions. Atmospheric Environment, 15, 2293.

Gotaas, Y., 1982, Vertical distribution of sulphur in the atmosphere in a case of long range transport and the rate of transformation to sulphate. Atmospheric Environment, 16, 1043.

Granat, L. and Johansson, C., 1983, Dry deposition of SO_2 and NO_x in winter. Atmospheric Environment 17, 191.

Gravenhorst, G., Böttger, A., 1982, Field measurements of NO and NO_2 fluxes to and from the ground. In: "Acid deposition" (S. Beilke and A.J. Elshout, eds.) p. 172, D. Reidel, Dordrecht, 1983.

Gravenhorst, G., Höfken, K.D. and Georgii, H.W., 1983, Acidic input to a beech and spruce forest. In: "Acid Deposition" (S. Beilke and A.J. Elshout, eds.), p. 155. Reidel, Dordrecht, 1983.

Grennfelt, P., Bengtson, C., Skärby, L., 1983, Dry deposition of nitrogen dioxide to scots pine needles. In: "Precipitation scavenging, dry deposition and resuspension", (H.R. Pruppacher et al., eds.) vol. 2, p. 753. Elsevier Science Publ. Co. New York, 1983.

Grosjean, D., 1983, Distribution of atmospheric nitrogeneous pollutants at a Los Angeles area smog receptor site. Environ. Sci. Technol., 17, 13.

Guesten, H., 1983, Photocatalytic degradation of atmospheric pollutants on the surface of metal oxides. Paper presented at the Advanced Study Institute on Chemistry of Multiphase Atmospheric Systems, Corfu, 1983.

Guicherit, R., Hout, K.D. van den, Huygen, C., Duuren, H. van, Römer, F.G. and Viljeer, J.W., 1980, Conversion rate of nitrogen oxides in polluted atmosphere, Proceedings of the 11th NATO-CCMS International Technical Meeting on Air Pollution Modelling and its Applications, 1980.

Hales, J.M., 1978, Wet removal of sulphur compounds from the atmosphere. Atmospheric Environment 12, 389.

Hales, J.M., and Dana, M.T., 1979a, Regional scale deposition of sulphur dioxide by precipitation scavenging. Atmospheric Environment 13, 1121.

Hales, J.M., and Dana, M.T., 1979b, Precipitation scavenging of urban pollutants by convective storm systems. Journal of Applied Meteorology, 18, 294-316.

Hales, J.M. and Sutter, S.L., 1973, Solubility of sulphur dioxide in water at low concentrations. Atmospheric Environment 7, 997.

Hanabusa, M., et al., 1977, Pulsewidth dependence of ozone interference in the laser fluorescence measurements of OH in the atmosphere. J. Chem. Phys., 66, 2118.

Hard, T.M., O'Brien, R.J., Chan, C.Y., Mehrabzadeh, A.A., 1984, Tropospheric free radical determination by FAGE. Environmental Science and Technology, 18, 768.

Hegg, D.A., Hobbs, P.V., Radke, L.F. and Harrison, H., 1977, Reactions of ozone and nitrogen oxides in power plant plumes. Atmospheric Environment, 11, 521.

Hegg, D.A., Hobbs, P.V. and Radke, L.F., 1984a, Measurements of the scavenging of sulphate and nitrate in clouds. Atmospheric Environment 18, 1939.

Hegg, D.A., Rutledge, S.A. and Hobbs, P.V., 1984b, A numerical model for sulphur chemistry in warm-frontal rainbands. Journal of Geophysical Research, 89, D5, 7133.

Hegg, D.A. and Hobbs, P.V., 1978, Oxidation of sulphur dioxide in aqueous systems with particular reference to the atmosphere. Atmospheric Environment, 12, 241.

Hegg, D.A. and Hobbs, P.V., 1980, Measurements of gas-to-particle conversion in the plumes from five coal-fired electric power plants. Atmospheric Environment 14, 99.

Hegg, D.A. and Hobbs, P.V., 1983, Preliminary measurements on the scavenging of sulphate and nitrate by clouds. In: "Precipitation scavenging, dry deposition and resuspension" (Pruppacher et al., (eds.)), Elsevier Science Publ., New York, 1983.

Heidema, L.F., and Verhagen, H.L.M., 1983, De droge depositie van luchtverontreiniging (Dry deposition of air pollution). TNO-report IMG G 1052 (in Dutch) TNO, Delft.

Heikes, B.G., and Thompson, A.M., 1983, Effects on heterogeneous processes on NO_3, HNO_2 and HNO_3 chemistry in the troposphere. Journal of Geophysical Research, 88, C15, 10883.

Heikes, B.G., and Thompson, A.M., 1984, Correction to "Effects of heterogeneous processes on NO_3, HONO, and HNO_3 chemistry in the troposphere". Journal of Geophysical Research, 89, D7, 11829.

Hicks, B.B., et al., 1980, Critique of methods to measure dry deposition, workshop summary, October 1980. Rapport EPA-600/9-80-050 NTIS PB81-126443.

Hicks, B.B., and Slinn, W.G.N., 1983, Surface fluxes of small particles. Draft report 1983. Atmospheric Turbulence and Diffusion Laboratory, Oak Ridge, USA.

Hicks, B.B., Wesely, M.L., Coulter, R.L., Hart, R.L., Durham, J.L., Speer, R.E., and Stedman, D.H., 1983, An experimental study of sulphur deposition to grassland. In: "Precipitation Scavenging, Dry Deposition and Resuspension" (H.R. Pruppacher et al., eds.) Vol. 2, p. 933.

Hicks, B.B., Wesely, M.L., Durham, J.L., and Brown, M.A., 1982, Some direct measurements of atmospheric sulphur fluxes over a pine plantation. Atmospheric Environment 16, 2899.

Hill, A.C., 1971, Vegetation: a sink for atmospheric pollutants. J. Air Pollut. Contr. Ass., 21, 341.

Hoffmann, M.R., 1983, Aqueous phase photo-assisted oxidation of S(IV) by oxygen on hematite (α-Fe_2O_3/s) at low pH. Paper presented at the Advanced Study Institute on Chemistry of Multiphase Atmospheric Systems, Corfu, 1983.

Höfken, K.D., Meixner, F.X. and Ehhalt, D.H., 1983, Deposition of atmospheric trace constituents onto different natural surfaces. In: "Precipitation Scavenging, Dry Deposition and Resuspension" (H.R. Pruppacher et al., eds.), Vol. 2, p. 825.

Högström, U., 1974, Wet fallout of sulphurous pollutants emitted from a city during rain or snow. Atmospheric Environment 9, 1291.

Holdren, M.W., Spicer, C.W., and Hales, J.M., 1984, Peroxyacetyl nitrate solubility and decomposition rate in acidic water. Atmospheric Environment 18, 1171.

Homolya, J.B. and Fortune, C.R., 1978, The measurement of the sulphuric acid and sulphate content of particulate matter resulting from the combustion of coal and oil. Atmospheric Environment, 12, 2511.

Horst, T.W., 1977, A surface depletion model for deposition from a gaussian plume. Atmospheric Environment, 11, 41.

Horst, T.W., 1983b, A correction to the Gaussian source-depletion model. In: "Precipitation scavenging, dry deposition and resuspension" (eds. Pruppacher et al.), vol. 2, p. 1205, Elsevier Science Publishing Co., New York, 1983.

Horst, T.W., Doran, J.C., and Nickola, P.W., (1983a), Dual tracer measurements of plume depletion. In: "Precipitation Scavenging. Dry Deposition and Resuspension" (H.R. Pruppacher et al., eds.) Vol. 2, p. 1027, Elsevier, New York, 1983.

Hout, K.D. van den, and Dop, H. van, 1985, Chapter 2 of this book.

Hov, Ø., and Isaksen, I.S.A., 1979, Hydroxyl and peroxy radicals in polluted tropospheric air. Geophys. Res. Lett., 6, 219.

Hübler, G., Perner, D., Platt, U., Tönnissen, A. and Ehhalt, D.H., 1984, Groundlevel OH radical concentration: New measurements by optical absorption. Journal of Geophysical Research 89, D1, 1309.

Huebert, B.J., Fehsenfeld, F.C., Norton, R.B. and Albritton, D., 1983, The scavenging of nitric acid vapor by snow. In: Precipitation scavenging, dry deposition and resuspension (Pruppacher et al., (eds.)), Elsevier Science Publ., New York, 1983.

Huebert, B.J., Wesely, M.L. and Stedman, D.H., 1982, Discussion on Wesely et al. (1982), Atmospheric Environment 16, 1600.

Huebert, B.J., 1983, Measurements of the dry-deposition flux of nitric acid vapor to grasslands and forest. In: "Precipitation scavenging, dry deposition and resuspension" (H.R. Pruppacher et al., eds.) vol. 2, p. 785. Elsevier Science Publ. Co. New York, 1983.

Huebert, B.J., Robert, C.H., 1985, The dry deposition of nitric acid to grass. J. Geophys. Res. 90, D1, 2085.

Husar, R.B., Patterson, D.E., Husar, J.D. and Gillani, N.V., 1978, Sulphur budget of a power plant plume. Atmospheric Environment, 12, 549.

Ibrahim, M., Barrie, L.A. and Fanaki, F., 1983, An experimental and theoretical investigation of the dry deposition of particles to snow, pine trees and artificial collectors. Atmospheric Environment 17, 781.

ISSA, 1978, International Symposium on Sulphur in the Atmosphere, Dubrovnik, 1977. Proceedings: Atmospheric Environment, 12, no. 1, 2, 3.

Johansson, C., Richter, A. and Granat, L., 1983, Dry deposition on coniferous forest of SO_2 at ppb levels. In: "Precipitation Scavenging, Dry Deposition and Resuspension" (H.R. Pruppacher et al., eds.) vol. 2, p. 775, Elsevier, New York, 1983.

Jonas, R., 1984, Deposition of aerosols in waste air plumes and filtering effects of meadows. Eleventh annual conference of the Association for Aerosol Research. J. Aerosol Sci. 15, 405.

Jones, C.L., and Seinfeld, J.H., 1983, The oxidation of NO_2 to nitrate, day and night. Atmospheric Environment 17, 2370.

Judeikis, H.S., 1978, Laboratory studies of heterogeneous reactions of SO_2. Atmospheric Environment, 12, 1633.

Judeikis, H.S., and Stewart, T.B., 1976, Laboratory measurements of SO_2 deposition velocities on selected building materials and soils. Atmosph. Environm., 10, 769.

Judeikis, H.S., and Wren, A.G., 1978, Laboratory measurements of NO and NO_2 depositions onto soil and cement surfaces. Atmospheric Environment, 12, 2315.

Junge, C.E., 1963, "Air Chemistry and Radioactivity". Academic Press, New York, 1963.

Kasting, J.F., 1980, Determination of deposition velocities for O_3, NO_x, and HNO_3 by the gradient method. In: the CHON photochemistry of the troposphere, p. 161. NCAR/CQ-7+1980-ASP, Boulder, 1980.

Katen, P.C., and Hubbe, J.M.,1983, Size-resolved measurements by eddy correlation of the dry deposition velocity of atmospheric aerosol particles. In: "Precipitation Scavenging, Dry Deposition and Resuspension" (H.R. Pruppacher et al., eds.) Vol. 2, p. 953.

Katen, P.C., and Hubbe, 1985, An evaluation of opticle particle counter measurements of the dry deposition of atmospheric aerosol particles. J. Geophys. Res. 90, D1 2145.

Kessler, C., 1979, UV-spektroskopische Bestimmung der trockenen Deposition von SO_2 and NO_2 mittels Gradientenmethode, (Diplomarbeit), Inst. f. atmosphärische Chemie, KFA Jülich.

Kewley, D.J., 1978, Atmospheric dispersion of a chemically reacting plume. Atmospheric Environment, 12, 1895.

Law, R.M., Mansfield, T.A., 1982, Oxides of nitrogen and the greenhouse atmosphere. In: "Effects of gaseous air pollution in agriculture and horticulture" (M.H. Unsworth and D.P. Ormrod, eds.) Butterworths, London.

Lazrus, A.L., Haagenson, P.L., Kok, G.L., Huebert, B.J., Kreitzberg, C.W., Likens, G.E., Mohnen, V.A., Wilson, W.E., and Winchester, J.W., 1983, Acidity in air and water in a case of warm frontal precipitation. Atmospheric Environment, 17, 581.

Leaitch, W.R., Strapp, J.W., Wiebe, H.A. and Isaac, G.A., 1983, Measurements of scavenging and transformation of aerosol inside cumulus, In: "Precipitation scavenging, dry deposition and resuspension" (Pruppacher et al., (eds.)), Elsevier Science Publ., New York, 1983.

Lee, Y.N., and Schwartz, S.E., 1981, Reaction kinetics of nitrogen dioxide with water at low partial pressure. J. Phys. Chem., 85, 840.

Legg, B.J. and Price, R.I., 1980, The contribution of sedimentation to aerosol deposition to vegetation with a large leaf area index. Atmospheric Environment 14, 305.

Levine, S.Z. and Schwartz, S.E., 1982, In-cloud and below-cloud scavenging of nitric acid vapor. Atmospheric Environment, 16, 1725.

Levy, A., Drewer, D.R., and Hales, J.M., 1976, SO_2 oxidation in plumes. A review and assessment of relevant mechanistic and rate studies. EPA 450/3-76-022.

Liberti, A., Brocco, D. and Possanzini, M., 1978, Adsorption and oxidation of sulphur dioxide on particles. Atmospheric Environment 12, 255.

Lindberg, S.E., 1982, Factors influencing trace metal, sulphate and hydrogen ion concentrations in air. Atmospheric Environment 16, 1701.

Lindberg, S.E., Lovett, G.M., 1983, Application of surrogate surface and leaf extraction methods to estimation of dry deposition to plant canopies. In: "Precipitation Scavenging, Dry Deposition, and Resuspension" (H.R. Pruppacher et al., eds.) Vol. 2, p. 837 Elsevier, New York, 1983.

Lindberg, S.E., and Lovett, G.M., 1985, Field measurements of particle dry deposition rates to foliage and inert surfaces in a forest canopy. Environ. Sci. Technol. 19, 238.

Little, P., and Wiffen, R.D., 1977, Emission and deposition of petrol engine exhaust Pb, I. Deposition of exhaust Pb to plant and soil surfaces. Atmospheric Environment 11, 437.

Logan, J.A., 1983, Nitrogen oxides in the troposphere: Global and regional budgets. Journal of Geophysical Research, 88, C15, 10785.

Lusis, M.A., 1978, Mathematical modelling of dispersion and chemical reactions in a plume - oxidation of NO to NO_2 in the plume of a power plant. Atmospheric Environment 12, 1231.

Lusis, M.A., Anlauf, K.G., Barrie, L.A., and Wiebe, H.A., 1978, Plume chemistry studies at a northern Alberta power plant. Atmospheric Environment, 12, 2429.

Lusis, M.A., and Wiebe, H.A., 1976, The rate of oxidation of SO_2 in the plume of a nickel smelter plant. Atmospheric Environment, 10, 793.

Makhon'ko, K.P., 1967, Simplified theoretical notion of contaminant removal by precipitation from the atmosphere, Tellus, 19, 467.

Martin, A., 1984, Estimated washout coefficients for sulphur dioxide, nitric oxide, nitrogen dioxide and ozone. Atmosperic Environment 18, 1955.

Martin, L.R., 1983, Kinetic studies of sulphite oxidation in aqueous solutions. In: "Acid Precipitation: SO_2, NO and NO_2 oxidation mechanisms: Atmospheric Considerations". Ann Arbor Scientific Publications, in press.

Maul, P.R., 1978, Preliminary estimates of the washout coefficient for sulphur dioxide using data from an East Midlands ground level monitoring network. Atmospheric Environment 12, 2515.

McLaren, E., 1983, Oxidation of SO_2 on surfaces of certain carbon particulates. Paper presented at the Advanced Study Institute on Chemistry of Multiphase Atmospheric Systems, Corfu, 1983.

McMahon, T.A., and Denison, P.J., 1979, Empirical atmospheric deposition parameters - A survey. Atmospheric Environment 13, 571.

McMurry, P.H. Rader, D.J. and Stith, J.L., 1981, Studies of aerosol formation in power plant plumes - I. Growth laws for secondary aerosols in power plant plumes: implications for chemical conversion mechanisms. Atmospheric Environment 15, 2315.

McMurry, P.H. and Wilson, J.C., 1983, Droplet phase (heterogeneous) and gas phase (homogeneous) contributions to secondary ambient aerosol formation as functions of relative humidity. Paper presented at the Advanced Study Institute on Chemistry of Multiphase Atmospheric Systems, Corfu, 1983.

Meagher, J.F., Stockburger, L., Bailey, E.M. and Huff, O., 1978, The oxidation of sulphur dioxide to sulphate aerosol in the plume of a coal-fired power plant. Atmospheric Environment, 12, 2197.

Meagher, J.F., Stockburger, L., Bonanno, R.J. and Luria, M., 1981, Cross-sectional studies of plumes from a partially SO_2-scrubbed power plant. Atmospheric Environment 15, 2263.

Meagher, J.F., and Olszyna, K.J., 1984, The effect of SO_2 gas phase oxidation on hydroxyl smog chemistry. Atmospheric Environment 18, 2095.

Melo, O.T., Lusis, M.A., and Stevens, R.D.S., 1978, Mathematical modelling of dispersion and chemical reactions in a plume - oxidation of NO to NO_2 in the plume of a power plant. Atmospheric Environment, 12, 1231.

Middleton, P., Kiang, C.S. and Mohnen, V.A., 1980, Theoretical estimates of the relative importance of various urban sulphate aerosol production mechanism. Atmospheric Environment, 14, 463.

Miller, D.F., Alkezweeny, A.J., Hales, J.M. and Lee, R.N., 1978, Ozone formation related to power plant emissions. Science 202, 1186.

Milne, J.W., Roberts, D.B. and Williams, D.J., 1979, The dry deposition of sulphur dioxide - field measurements with a stirred chamber. Atmospheric Environment 13, 373.

Möller, D., 1980, Kinetic model of atmospheric SO_2 oxidation based on published data. Atmospheric Environment, 14, 1067.

Nester, K., Verenkotte, H., Dlugi, R., Bunz, H. and Jordan, S., 1984, Modelling of photochemical and heterogeneous sulphate aerosol formation in plumes. J. Aerosol Sci. 15, 399.

Nestlen, M., Münnich, K.O., Zenger, A., Flothmann, D. and Platt, U., 1982, Austausch von Luftverunreinigungen an der Grenzfläche Atmosphäre/Erdoberfläche (trockene Deposition), Zwischenbericht zum Teilprojekt 2: Experimentelle Arbeiten. Battelle-Institut Frankfurt, report BleV-R-64.284-4.

Neumann, H.H., and Hartog, G. den, 1985, Eddy correlation measurements of atmospheric fluxes of ozone, sulphur and particulates during the Champaign intercomparison study. J. Geophys. Res. 90, D1, 2097.

Newman, L., Forrest, J. and Manowitz, B., 1975, The application of an isotopic ratio technique to a study of the atmospheric oxidation of sulphur dioxide in the plume from an oil-fired power plant. Atmospheric Environment, 9, 959.

Newman, L., 1981, Atmospheric oxidation of sulphur dioxide: a review as viewed from power plant and smelter plume studies. Atmospheric Environment 15, 2231.

Niemann, B.L., 1983, Scavenging ratios for exeptional wet sulphate episodes in eastern North America from three event networks, In: "Precipitation scavenging, dry deposition and resuspension" (Pruppacher et al., (eds.)), Elsevier Science Publ., New York, 1983.

Novakov, T., and Chang, S.G., 1976, Catalytic oxidation of SO_2 on carbon particles. AICHE Symposium Series, 72, 255.

O'Brien, E.E., et al., 1976a, Chemically reactive turbulent plumes. Dep. Mech. Eng. State Univ. New York. Report 1976, BNL - 21483.

O'Brien, E.E., et al., 1976b, Chemically reactive turbulent plumes. Proceedings third symposium on atmospheric turbulence, diffusion and air quality. American meteorological society, Boston, MA 1976 A 160.

Onderdelinden, D., Jaarsveld, J.A. van, and Egmond, N.D. van, 1984, Bepaling van de depositie van zwavelverbindingen in Nederland (Determination of the deposition of sulphur compounds in the Netherlands. In Dutch) Report RIVM 842017001. Rijksinstituut voor Volksgezondheid en Milieuhygiëne, Bilthoven, the Netherlands.

Overton, Jr., J.H., 1978, Discussion on: Oxidation of sulphur dioxide in the Sudbury smelter plume. Atmospheric Environment, 12, 2029.

Payrissat, M., and Beilke, S., 1975, Laboratory measurements of the uptake of sulphur dioxide by different European soils. Atmospheric Environment, 9, 211.

Pena, J.A., Pena, R.G. de, Bowersox, V.C., and Takacs, J.F., 1982, SO_2 content in precipitation and its relationship with surface concentrations of SO_2 in the air. Atmospheric Environment 16, 1711.

Perner, D., Ehhalt, D.H., Plätz, H.W., Platt, U., Röth, E.P. and Volz, A., 1976, OH radicals in the lower troposphere. Geophys. Res. Lett., 3, 466.

Persson, C., 1983, Plume model for nitrogen oxides. 14th I.T.M. on Air Pollution Modelling and its Applications, Copenhagen (1983).

Platt, U., Perner, D., Harris, G.W., Winer, A.M. and Pitts jr., J.N., 1980a, Observation of nitrous acid in an urban atmosphere by differential optical absorption. Nature, 285, 312.

Platt, U., and Perner, D., 1980b, Direct measurements of atmospheric CH_2O, HNO_2, O_3, NO_2, and SO_2 by differential optical absorption in the near UV. Journal of Geophysical Research, 85, C12, 7453.

Platt, U., Perner, D., Winer, A.M., Harris, G.W. and Pitts jr, J.N., 1980c, Detection of NO_3 in the polluted troposphere by differential optical absorption. Geophys. Res. Lett., 7, 89.

Platt, U., Winer, A.M., Biermann, H.W., Atkinson, R., and Pitts, J.N., 1984, Measurement of nitrate radical concentrations in continental air. Environ. Sci. Technol., 18, 365.

Pruppacher, H.R., Semonin, R.G., and Slinn, W.G.N., 1983, "Precipitation Scavenging, Dry Deposition and Resuspension. Volume 2, Dry deposition and Resuspension". Elsevier, New York, 1983.

Radke, L.F., 1983, Preliminary measurements of the size distribution of cloud interstitial aerosol. In: "Precipitation scavenging, dry deposition and resuspension" (Pruppacher et al., (eds.)), Elsevier Science Publ., New York, 1983.

Raynor, G.S. and Hayes, J.V., 1983, Differential rain and snow scavenging efficiency implied by ionic concentration differences in winter precipitation, In: "Precipitation scavenging, dry deposition and resuspension" (Pruppacher et al., (eds.)), Elsevier Science Publ., New York, 1983.

Reynolds, S.D., et al., 1979, An introduction to the SAI airshed model and its usage. System Applications Inc. San Rafael, Cal., Report EF 78-53 R4 - EF 79-31.

Richards, L.W., 1983, Comments on the oxidation of NO_2 to nitrate, day and night. Atmospheric Environment, 17, 397.

Roberts, B.R., 1974, Foliar sorption of atmospheric sulphur dioxide by woody plants. Environ. Pollut. 7, 133.

Rogers, H.H., Jeffries, H.E., Witherspoon, A.M., 1979, Measuring air pollutant uptake by plants: nitrogen dioxide. J. Environmental Qual. vol. 8, 551.

Schmidt, J., Eastman, J., and Sievering, H., 1983, Influence of relative humidity and sea salt nuclei on the eddy flux determination of small particle dry deposition over the sea. In: "Precipitation Scavenging, Dry Deposition and Resuspension" (H.R. Pruppacher et al., eds.) Vol. 2, p. 1233.

Schwartz, S.E., Daum, P.H., Hjelmfelt, M.R. and Newman, L., 1983, Cloud water acidity measurements and formation mechanisms - experimental design. In: "Precipitation scavenging, dry deposition and resuspension" (Pruppacher et al., (eds.)), Elsevier Science Publ., New York, 1983.

Schwartz, S.E. and Freiberg, J.E., 1981, Mass-transport limitation to the rate of reaction of gases in liquid droplets: Application to oxidation of SO_2 in aqueous solutions. Atmospheric Environment 15, 1129.

Schwartz, S.E., and Newman, L., 1978, Processes limiting the oxidation of sulphur dioxide in stack plumes. Environmental Science and Technology, 12, 67.

Schwela, D., 1977, Die trockene Deposition gasförmiger Luftverunreinigungen. Schriftenreihe der Landesanstalt für Immissionsschutz, Heft 42, 46.

Scott, B.C., 1981, J. Appl. Meteorol., 20, 619.

Scott, B.C., 1982a, Theoretical estimates of the scavenging coefficient for soluble aerosol particles as a function of precipitation type, rate and altitude. Atmospheric Environment 16, 1753.

Scott, B.C., 1982b, Prediction of in-cloud conversion rates of SO_2 to SO_4 based upon a simple chemical and kinematic storm model. Atmospheric Environment 16, 1735.

Scott, B.C. and Laulainen, N.S., 1979, On the concentration of sulphate in precipitation. J. Appl. Met. 18, 138.

Sehmel, G.A., 1980, Particle and gas dry deposition: a review. Atmosph. Environment, 14, 983.

Sehmel, G.A., 1983, Particle dry deposition measurements with dual tracers in field experiments. In: "Precipitation Scavenging, Dry Deposition and Resuspension" (H.R. Prappacher et al., eds.) Vol. 2, p. 1013, Elsevier, New York, 1983.

Sehmel, G.A., Sutter, S.L., 1974, Particle deposition rates on a water surface as a function of particle diameter and air velocity. J. de Recherche Atmos. 8, 911.

Sehmel, G.A., and Hodgson, W.H., 1980, A model for predicting dry deposition of particles and gases to environmental surfaces. A.I.Ch.E. Symposium Series, Vol. 76, no. 196, 218.

Seigneur, C., and Saxena, P., 1984, A study of atmospheric acid formation in different environments. Atmospheric Environment, 18, 2109.

Seinfeld, J.H., and Wilson, K.R., 1977, International conference on oxidants 1976 - analysis of evidence and viewpoints. Part VI, The issue of air quality simulation model utility. EPA - 600/3 - 77 - 118 (1977).

Selzer, P.M., and Wang, C.C., 1979, Quenching rates and fluorescence efficiency in the $A^2\varepsilon^+$ state of OH. J. Chem. Phys., 71, 3786.

Shu, W.R., Lamb, R.G. and Seinfeld, J.H., 1978, A model of second-order chemical reactions in turbulent fluid - part II, Application to atmospheric plumes. Atmospheric Environment 12, 1695.

Sickles, J.E., Bach, W.D. and Spiller, L.L., 1983, Comparison of several techniques for determining dry deposition flux. In: "Precipitation Scavenging, Dry Deposition and Resuspension" (H.R. Pruppacher et al., eds.), Vol. 2, p. 979, Elsevier, New York, 1983.

Sievering, H., 1981, Profile measurements of particle mass transfer at the air-water interface. Atmospheric Environment, 15, 123.

Sievering, H., 1982, Profile measurements of particle dry deposition velocity at an air-land interface. Atmospheric Environment 16, 301.

Sievering, H., 1983, Eddy flux and profile measurements of small-particle dry deposition velocity at the Boulder Atmospheric Observatory. In: "Precipitation Scavenging, Dry Deposition and Resuspension" (H.R. Pruppacher et al., eds.) Vol. 2, p. 963. Elsevier, New York, 1983.

Sievering, H., Eastman, J. and Schmidt, J.A., 1982, Air-sea particle exchange at a nearshore oceanic site. J. Geophys. Res. 87, 11027.

Sievering, H. Van Valin, C.C., Barrett, E.W., and Pueschel, R.F., 1984, Cloud scavenging of aerosol sulphur: Two case studies. Atmospheric Environment 18, 2685.

Sievering, H., and Pueschel, R., 1982, Impact of particle characterisation in carfounding reported particle deposition velocities. Atmospheric Environment 16, 359.

Slinn, W.G.N., 1982a, Predictions for particle deposition to vegetative canopies. Atmospheric Environment 16, 1785.

Slinn, W.G.N., 1982b, Precipitation scavenging, In: "Atmospheric Sciences and Power Production", D. Randerson (editor); US DOE Technical Information Center, Oak Ridge, 1982.

Slinn, W.G.N., 1983a, A potpourri of deposition and resuspension questions. In: "Precipitation Scavenging, Dry Deposition and Resuspension" (H.R. Pruppacher et al., eds.) Vol. 2, p. 1361, Elsevier, New York, 1983.

Slinn, W.G.N., 1983b, Air-to-sea transfer of particles. In: "Air-sea Exchange of Gases and Particles", (P.S. Liss, W.G.N. Slinn, eds.) p. 299. D. Reidel, 1983.

Slinn, W.G.N., Hasse, L., Hicks, B.B., Hogan, A.W., Lal, D., Liss, P.S., Munnick, K.O., Sehmel, G.A. and Vittori, O., 1978, Some aspects of the transfer of atmospheric trace constituents past the air-sea interface. Review paper. Atmospheric Environment 12, 2055.

Slinn, S.A., and Slinn, W.G.N., 1980, Predictions for particle deposition on natural waters. Atmospheric Environment 14, 1013.

Smith, T.B., et al., 1975, The chemistry, dispersion and transport of air pollutants emitted from fossil fuel power plants in California. U.S. NTIS rep. 1975 PB-254449.

Spicer, C.W., 1977, Photochemical atmospheric pollutants derived from nitrogen oxides. Atmospheric Environment, 11, 1089.

Spicer, C.W., 1980, The rate of NO_x reaction in transported urban air. In: "Atmospheric Pollution 1980", M.M. Benarie, ed., Elsevier Sci. Pub. Cie, Amsterdam 1980, p. 181.

Spicer, C.W., 1982, Nitrogen oxide reactions in the urban plume of Boston, Science, 215, 1095.

Spicer, C.W., and Miller, D. F. 1976, Nitrogen balance in smog chamber studies. J. Air Pollut. Control Ass., 26, 45.

Sprugel, D.G., and Miller, J.E., 1979, A field estimate of SO_2 deposition velocities to rapidly growing soybeans. Water Air and Soil Pollution 12, 233.

Stockwell, W.A., and Calvert, J.G., 1983, The mechanism of the HO-SO_2 reaction. Atmospheric Environment 17, 2231.

Sverdrup, G.M., and Hov, Ø., 1984, Modelling study of the potential importance of heterogeneous surface reactions for NO_x transformations in plumes. Atmospheric Environment 18, 2753.

Taylor, G.E., McLaughlin, S.B., Shriner, D.S., and Selvidge, W.J., 1983, The flux of sulphur-containing gases to vegetation. Atmospheric Environment 17, 789.

Wang, C.C., Davis, L.I., Wu, C.H., Japar, S., Niki, H. and Weinstock, S., 1975, Hydroxyl radical concentrations measured in ambient air. Science, 189, 797.

Wedding, J.B., Carlson, R.W., Stukel, J.J., en Bazzaz, F.A., 1977, Aerosol deposition on plant leaves. Water, Air and Soil Pollution 7, 545.

Weinstock, B., Niki, H. and Chang, T.Y., 1980, Chemical factors affecting the hydroxyl radical concentration in the troposphere. "Advances in Environmental Science and Technology", 10, 221.

Wells, A.C. and Chamberlain, A.C., 1967, Transport of small particles to vertical surfaces. Brit. J. Appl. Phys. 18, 1793.

Wesely, M.L., Hicks, B.B., Dannevik, W.P., Frisella, S. and Husar, R.B., 1977, An eddy correlation measurement of particulate deposition from the atmosphere. Atmospheric Environment, 11, 561.

Wesely, M.L., and Hicks, B.B., 1979, Dry deposition and emission of small particles at the surface of the earth. Preprint Vol. Fourth Symposium on Turbulence, Diffusion and Air Pollution, Reno, NV, USA. Am. Meteor.Soc. Boston MA pp. 510-513.

Wesely, M.L., Cook, D.R. and Hart, R.L., 1985, Measurements and parametrisation of particulate sulphur dry deposition over grass. J. Geophys. Res., 90, D1, 2131.

Wesely, M.L., Eastman, J.A., Stedman, D.H., Yalvac, E.D., 1982, An eddy correlation measurement of NO_2 flux to vegetation and comparison to O_3 flux. Atmospheric Environment 16, 815.

Wesely, M.L., Cook, D.R., Hart, R.L., Hicks, B.B., Durham, G.L., Speer, R.E., Stedman, D.H. and Tropp, R.J., 1983a, Eddy correlation measurements of the dry deposition of particulate sulphur and submicron particles. In: "Precipitation Scavenging, Dry Deposition and Resuspension" (H.R. Pruppacher et al., eds.), Vol. 2, p. 943, Elsevier, New York, 1983.

Wesely, M.L., Cook, D.R., and Hart, R.L., 1983b, Fluxes of gases and particles above a deciduous forest in winter time. Boundary Layer Meteorology, 27, 237.

Whelpdale, D.M., 1981, Rep. EMEP/CCC 5/81. Norwegian Institute for Air Research Lillestrøm, Norway.

Whitby, K.T., 1978, The physical characteristics of sulphur aerosols. Atmospheric Environment 12, 135.

Whitby, K.T., Cantrell, B.K., and Kittelson, D.B., 1978, Nuclei formation rates in a coal-fired power plant plume. Atmospheric Environment, 12, 313.

White, W.H., 1977, NO_x - O_3 photochemistry in power plant plumes: comparison of theory with observation. Environmental Science and Technology, 11, 995.

Wiel, H.J., van de, et al., 1977, Instrument performance. Rijksinstituut voor de Volksgezondheid, Bilthoven, 1977.

Williams, D.J., Carras, J.M., and Milne, J.W., 1981, The oxidation and long-range transport of sulphur dioxide in a remote region. Atmospheric Environment, 15, 2255.

Williams, R.M., 1982, A model for the dry deposition of particles to natural water surfaces. Atmospheric Environment 17, 1933.

Wilson, J.C., and McMurry, P.H., 1981, Studies of aerosol formation in power plant plumes - II. Secondary aerosol formation in the Navajo generating station plume. Atmospheric Environment 15, 2329.

Wilson, W.E., 1978, Sulphates in the atmosphere: a progress report on Project MISTT. Atmospheric Environment, 12, 537.
Wilson, W.E., 1981, Sulphate formation in point source plumes: a review of recent field studies. Atmospheric Environment 15, 2573.
Wilson, W.E., et al., 1976, Chemical reactions in power plant plumes. Prep. Pap. Natl. meeting Div. Environm. Chem.. Am. Chem. Soc., 16 (1), 60.
Wiman, B.L.B., and Ågren, G.I., 1985, Aerosol depletion and deposition in forests - a model analysis. Atmospheric Environment 19, 335.
Winer, A.M., Peters, J.W., Smith, J.P., Pitts jr., J.N., 1974, Response of commercial chemiluminescent NO-NO_2 analyzers to other nitrogen-containing compounds, Env. Sci. and Techn., 8, 1118.
Yoshizumi, K., and Hoshi, A., 1985, Size distribution of ammonium nitrate and sodium nitrate in atmospheric aerosols. Environ. Sci. Technol., 19, 258.
Zak, B.D., 1981, Lagrangian measurements of sulphur dioxide to sulphate conversion rates. Atmospheric Environment 15, 2583.

CHAPTER 4

COMPARISON BETWEEN FOUR DIFFERENT INTERREGIONAL AIR POLLUTION MODELS

W. Klug[1], P.J.H. Builtjes[2], H. van Dop[3], N.D. van Egmond[4], H. Glaab[1], D. Gömer[1], B.J. de Haan[3], K.D. v.d. Hout[2], N. Kesseboom[4], R. Röckle[1], C. Veldt[2]

[1] TH Darmstadt, Institut für Meteorologie
6100 Darmstadt, Hochschulstr. 1
Federal Republic of Germany

[2] MT-TNO, P.O. Box 342, 7300 AH Apeldoorn
The Netherlands

[3] KNMI, P.O. Box 201, 3730 AE De Bilt
The Netherlands

[4] RIV, P.O. Box 1, 3720 BA Bilthoven
The Netherlands

LIST OF SYMBOLS

I. Common symbols

C	= concentration
t	= time
U,V,W	= wind in x, y and z-direction
K_x,K_y,K_z	= eddy diffusion in x, y and z-direction
D_D	= dry deposition
D_W	= wet deposition
S	= sources or/and sinks
r	= distance (box centre ← measurement point)
P	= precipitation rate
T	= temperature

V_D = deposition velocity

L = Monin - Obukhov - Length

u_* = friction velocity

g = geopotential

v_g = geostrophic wind

SO_2 = sulphur dioxide

M = mass

h = height

k_c = chemical removal rate

q_v, q_z = standard deviations in y- and z-direction

II. additional THD-symbols

C_h = chemical removal

H_{max} = top of layer in which the diurnal temperature variatio[n] appears

V = wind vector $V = \begin{matrix} u \\ v \\ w \end{matrix}$

a,b,c_o = constants

r_a = resistance concerning windstress and stability

r_D = deposition resistance

r_e = resistance concerning surface roughness

III. additional KNMI-symbols

Z_i = mixing height

U_i = observed West-East component of the wind velocity at the i-th synoptic station

V_a = index a marks the interpolation or gridpoint

V_i = observed North-South component of the wind velocity at the i-th synoptic station

C_{ij} = covariance of the observations of U at the stations i and j

Z_o	= local roughness length
r_i	= distance between the observation and interpolation point
a	= "radius of influence" of the data
m	= number of observations
U_{go}	= surface level geostrophic wind
U_{gl}	= 850 mbar geostrophic wind
z_1, z_s	= levels where U_{go} and U_{gl} are determined
α_s, α_g	= weighting functions
C_1, C_2	= constants
γ	= lapse rate of the layer beyond the inversion
θ	= mixed-layer potential temperature
Δ	= inversion strength
$\overline{\theta W_o}$	= heat flux at the surface
$\overline{\theta W_i}$	= " " " " inversion heigth
K	= von Karman constant

IV <u>additional TNO-symbols</u>

SF	= surface field
P_i	= profile field, where i refers to the layer number
G	= mixing height
h_n	= night time mixing height
K_H	= horizontal diffusion
R	= photolysis rate const. for NO_2
U_r	= reference velocity
α_{jk}	= fraction of grid square j represented by land use
β	= a factor that adjusts the reference surface uptake velocity to that for land use k
V_{si}	= reference surface uptake velocity for pollutant i

V additional RIV-symbols

f	= fraction of concentration affecting ground level
K_{zm}	= "bulk-turbulent diffusivity"
$\sigma_{y,o}$	= initial standard deviation in y-direction
K_H	= time dependent apparent horizontal diffusivity
t_p	= puff-travel time
$V_g 4$	= surface layer gradient
WV	= windvelocity
$\bar{N}$	= cloudcover
m	= power law windprofile exponent

VI additional symbols which occur in Sub-chapter III

x_i	= x,y,z
u_i	= u,v,w
K_{xi}	= K_x, K_y, K_z
Q	= source strength
h_s	= source height
H	= top boundary height
N	= number of grid points
C_{1i}, C_{2i}	= concentrations of two fields at the i-th grid point

I. INTRODUCTION

The "Committee on the Challenges of Modern Society" (CCMS) has been active in air pollution research since 1969. It undertook three pilot studies on various aspects of air pollution. The last one, a pilot study on "Air Pollution Control Strategies and Impact Modelling" was completed in 1984. The study has been carried out by three panels: one concerned with emissions, a second with air quality modelling, and a third with environmental impact assessment. The air quality modelling panel was assigned the task of reviewing the state of the art in relation to interregional modelling (CCMS study 1981; an updated version is presented as chapter 2 of this volume) and on the removal and transformation processes with respect to SO_2 and NO_x (CCMS study, 1981a; an updated version is presented as chapter 3 of this volume). A third document contains the comparison of a number of interregional transport models. The results of this intercomparison, carried out at the TH Darmstadt by Klug and his coworkers, are presented in this chapter. The intercomparison study was sponsored in part by the Umweltbundesant in Berlin, Federal Republic of Germany and the Ministry of Housing, Physical Planning and Environment in Leidschendam, The Netherlands.

It was explicitly stated in the beginning that the intercomparison study should not be directed to finding out which is the best of the models. Quite apart from the fact that it is difficult to objectively define which are the criteria to find the best model. No effort was made in this study to achieve the best agreement between model results and measurements. Furthermore it also became clear that - since initial concentrations and boundary fluxes of SO_2 were not considered - the results could not be used to select a model which gave the best performance. The main emphasis therefore should be put on an intercomparison of the results resulting from the four different models. Despite this however, a comparison with observed data was made for the last phase of the project in spite of the deficiencies mentioned. Another purpose of the study was to find out how the different models deal with certain meteorological situations and how large are the differences. It should also be stated here that a point measurement cannot be used to compare with model results which usually are volume averaged concentrations over large volumes in the order of 10 km x 10 km x 50 m.

Before the comparison study was begun, an expert workshop was held at Friedrichshafen, Germany (7th-9th December 1981). This workshop had the task of defining the selection criteria for the models, and which criteria were considered to be suitable for intercomparison. In addition hereto a workplan for the intercomparison study had to be developed.

It was decided that the models to be compared should be episode models, i.e. models simulating concentrations and depositions for time periods in the order of from one hour duration to a few days. Model comparisons on long term averages had already been performed by our North American colleagues under the US-Canadian "Memorandum of Intent". It was also agreed upon to emphasize transport on a scale of approximately 500 km since studies on larger scales had already been undertaken and the merits of short range (urban) models have been extensively tested. Furthermore, the models should be tested for sulphur dioxide as the emitted pollutant.

Other selection criteria were:

- the time resolution should be smaller than three hours;
- the spatial resolution should be in the order of 10-25 km;
- the vertical atmospheric structure pertinent to the dispersion of air pollution should be resolved, i.e. no single-layer models should be included;
- the physical processes to be described by the model are advection, turbulent diffusion and dry and wet deposition;
- removal by chemical processes should be accounted for;
- the meteorological input data which the various models would require should be derived from routinely available meteorological data;
- all models should be able to process complex multiple source areas.

The foregoing requirements had already put a heavy constraint on the number of potentailly available models and had reduced their number to approximately ten. On request, four models finally became available for effecting the intercomparison. These were the KNMI mesoscale transport model, the RIV puff model, an enlarged version of the SAI urban air-shed model and the THD model. The participants making the intercomparison decided to test their models using one test region only, and for which a sufficient data base would be available.

The principal idea behind this study was that all models of different complexity had to be run on the same input data bases such as emission inventory and meteorological routine observations as performed by the Weather Services. The chemical transformations were applied similarly to all the models.

A suitable flat region for the comparison was found to be the Netherlands and the surrounding countries and through which the orography could be neglected in the models. Figure I.1 shows the area used for this study. Its dimensions are 350 x 500 km^2, and it is approximately situated between 3° and 8° eastern longitude and between 49° and 54° north in latitude. It comprises the western

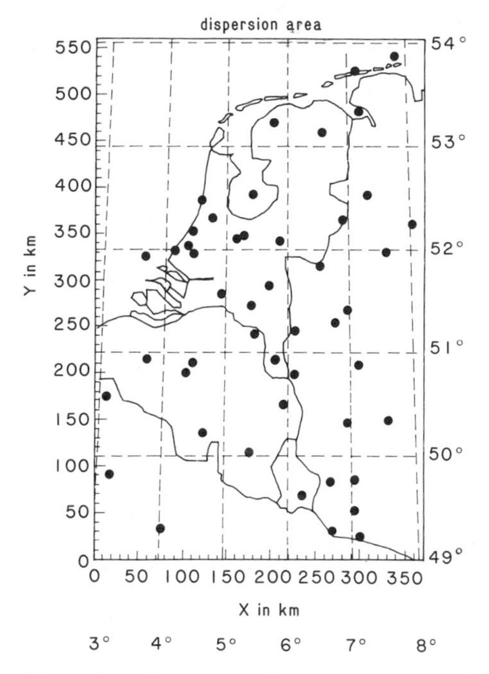

Fig. I.1 Area used for this study.
● Position of meteorological station

part of the Federal Republic of Germany, the Netherlands, Belgium, Luxemburg and a small part of North-West France.

This region includes:

(i) a dense network of routine meteorological observation stations,

(ii) an appropriate network of air-pollution monitoring stations,

(iii) an emissions inventory of the major pollutants which is relatively well-known when compared to most other areas of the same size.

Apart from this it contains Western Europe's main industrial areas: the Ruhr area, and the Rotterdam and Antwerp industrial areas; it is also one of the most densely populated areas in Europe.

Furthermore it was decided that flux across the boundaries, initial concentrations and background concentrations should not be considered in the calculations.

Three phases were distinguished between in the intercomparison. In phase I, a few simple numerical tests were designed with which model output could be compared with analytical solutions. In phase II the models were tested using realistic flow fields derived from the meteorological data. However, a hypothetical single point source was considered. This offered the possibility to study the differences in the processing of the meteorological fields by the four models.

Finally in phase III, three 48-hour case studies were undertaken, using the appropriate meteorology and emission inventory. Observed data were available in order to compare the model results with field measurements.

A summary of the models used will be given in the next section.

II. DESCRIPTION OF THE MODELS

In table II.1 an overview is given of the main features of the four models in a condensed manner. A comprehensive description of each model is given in the following sections.

II.a The THD (Technische Hochschule Darmstadt) - model

This model has been described in detail by Herrmann, 1978.

Basic features

The THD-model is a three dimensional Eulerian box model which describes the dispersion of a pollutant. For the solution of the following advection-diffusion equation

$$\frac{\partial c}{\partial t} = -u\frac{\partial c}{\partial x} - v\frac{\partial c}{\partial y} - w\frac{\partial c}{\partial z} + \frac{\partial}{\partial x}\left[k_x \frac{\partial c}{\partial x}\right] + \frac{\partial}{\partial y}\left[k_y \frac{\partial c}{\partial y}\right] +$$

$$\frac{\partial}{\partial y}\left[k_z \frac{\partial c}{\partial z}\right] - D_D - D_w - C_h + S \qquad \text{(II.1)1)}$$

a Eulerian explicit difference scheme is used. The advection part of the equation is solved by a time-split upstream differencing method which first treats the advection in x-direction and then, in a second step, the advection in y-direction. Afterwards vertical transport is calculated together with diffusion-, source-, deposition and chemistry terms. In this part of the equation, an upstream differencing scheme for the vertical advection and a centered differencing method for the diffusion terms is applied. The time-step

1) All symbols are defined in the list on page 150.

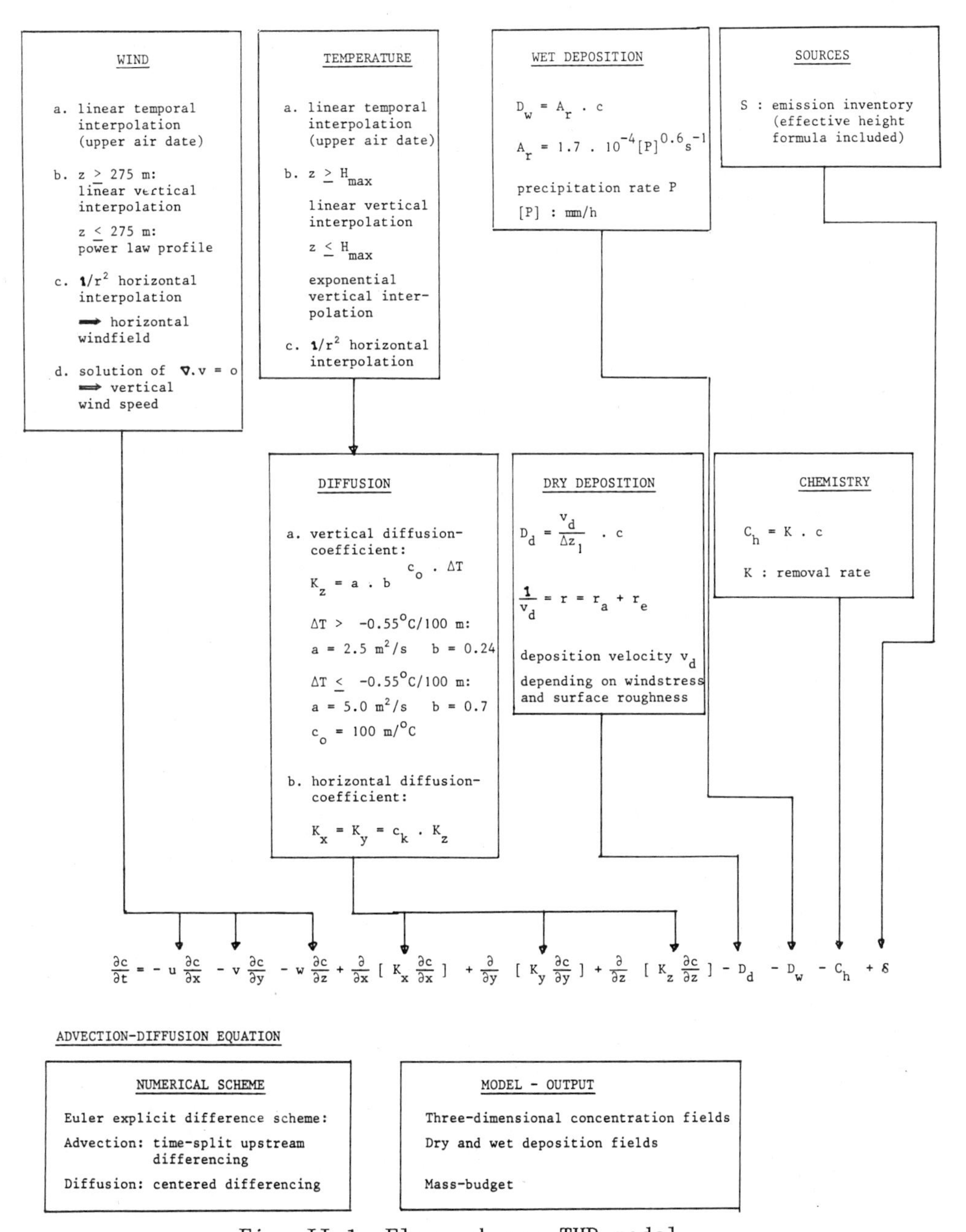

Fig. II.1 Flow scheme: THD-model.

Table II.1 Important Aspects of the Intercomparison.

	VERTICAL RESOLUTION USED IN THE STUDY	NUMERICAL METHOD	PLUME RISE	GRID SIZE USED IN THE STUDY	SURFACE WINDFIELD USED IN THE STUDY	VERTICAL DIFFUSION
THD GRID-MODEL	12 LAYERS	TIME-SPLIT UPSTREAM DIFFERENCE	MODIFIED BRIGGS-METHOD	10 x 10 KM^2	$1/r^2$-INTERPOLATION	BASED ON TEMPERATURE LAPSE RATE
KNMI GRID-MODEL	12 LAYERS	PSEUDO-SPECTRAL	BRIGGS-METHOD	20 x 20 KM^2*	INTERPOLATION BY COVARIANCES	BROST AND WYNGAARD
TNO (SAI) GRID-MODEL	5 LAYERS	SHASTA	BRIGGS-METHOD	10 x 10 KM^2	$1/r$-INTERPOLATION	DEARDORFF, BUSINGER AND ARYA
RIV PUFF-MODEL	2½ LAYERS*	ANALYTICAL TREATMENT	MODIFIED BRIGGS-METHOD	10 x 10 KM^2	NEGATIVE EXPONENTIAL WEIGHTING SCHEME	PASQUILL-VALUES

* no flexible parameter

Table II.1 Important Aspects of the Intercomparison. (continued)

MODEL	MIXING HEIGHT	HORIZONTAL DIFFUSION	DRY DEPOSITION	WET DEPOSITION	CONVERSION
THD	BASED ON VERTICAL DIFFUSION	PROPORTIONAL TO VERTICAL DIFFUSION	FUNCTION OF STABILITY AND SURFACE RESISTANCE	PROPORTIONAL TO RAIN INTENSITY	1% hr^{-1}
KNMI	TENNEKES AND NIEUWSTADT	NEGLECTED	"	"	"
TNO(SAI)	ESTIMATED FROM SURFACE DATA MAINLY**	50 M^2/S	"	"	FUNCTION OF PHOTOLYSIS RATE
RIV	ACOUSTIC SOUNDINGS**	FUNCTION OF LAGRANGIAN CORRELATION FUNCTION AND CROSS WIND TURBULENCE	"	NOT TREATED	1 % hr^{-1}

** values are input to the model

is variable and depends on wind velocity, diffusion coefficient, grid size and decay-terms. The dispersion area is divided into single boxes with a horizontal extension of 10 km and heights varying between 50 m and 300 m. In order to determine the concentration in the boxes, a mass-budget (Eq. II.1) is computed from the difference of the inflow and outflow of material through the boxboundaries. Also sources (emissions) and sinks (dry and wet deposition, linear chemical removal) are considered. This mass-budget is identical with the numerical solution of the diffusion equation (see flowdiagram Fig. II.1). The model can be used for a simulation covering several days in an area with a horizontal range of 100 km to 1000 km.

The emissions used in the model are treated in two different ways: For simulating elevated industrial sources an effective source height is computed and the emissions are located in the corresponding model layer. Sources situated near the surface, are emitting in the ground-layer of the model. The plume rise is simulated by a modified Briggs (1971) scheme.

The model produces hourly or half-hourly 3-dimensional concentrationfields of the pollutant. The values must be considered as mean concentrations of the box-volumes. Dry and wet deposition fields are routinely evaluated also. Additionally mass budgets such as, for example, loss caused by chemical removal processes and fluxes across the model boundaries can be computed.

Description of the windfield

For the set up of the horizontal wind and temperature fields, a $1/r^2$ interpolation (r = distance: boxcenter measurement point) is carried out. For the vertical scale, a power-law profile for the wind is used up to a height of 275 m. Above this height the winds at the box-centers are computed by linear interpolation of upper air data. For the temperature field the same procedure must be carried out except near the surface, where a modified method is used with respect to warming up and cooling down processes during day and night. The vertical mean wind velocities are computed using the divergence of the horizontal windfield.

The vertical and horizontal diffusion

The vertical eddy diffusion coefficients, necessary for the calculation of the turbulent fluxes, can be obtained by the use of the following equation

$$K_z = a \cdot b^{c_o \Delta T}$$

$\Delta T > -0.55°C/100$ m: $a = 2.5 \frac{m^2}{s}$; $b = 0.24$

$\Delta T \leq -0.55°C/100$ m: $a = 5.0 \frac{m^2}{s}$; $b = 0.7$

$c_o = 100$ m/°C

K_z depends on the lapse rate of the temperature.
The horizontal diffusion is proportional to the vertical diffusion.

Conversion, dry and wet deposition

The conversion process from SO_2 to SO_4 is dealt with as if it were a removal of the SO_2, with a removal rate of one percent per hour. The dry deposition is simulated by using a canopy resistance depending on the time of day and the windspeed. Wet deposition is coupled with the intensity of the local precipitation.

II.b The KNMI-model

The KNMI (= Royal Netherlands Meteorological Institute)-model is decribed in detail by Van Dop et al (1982/1983).

Basic features

The model is a Eulerian grid model. The basic formula solved is:

$$\frac{\partial c}{\partial t} + u\frac{\partial c}{\partial x} + v\frac{\partial c}{\partial y} = \frac{\partial}{\partial z}\left(k_z \frac{\partial c}{\partial z}\right) + S \qquad \text{(II.2)}$$

The model provides hourly concentrations in an area of about 400 x 400 km^2 and up to a height of 600 m during episodes of a few days.

For the numerical method, the fractional steps method is used.

The advection equation is solved by using the so-called "pseudo spectral method" (Gottlieb and Orszag, 1977), an efficient system with small numerical errors (see Van Dop et al 1982 and De Haan (1980)). The resulting first order ordinary differential equations are solved by a 4th order Runge Kutta Scheme (Gear, 1971). The (parabolic) diffusion equation is solved by applying a Crank-Nicholson scheme (Richtmeyer and Morton, 1967).

Surface sources are introduced at the centres of the lowest grid cells (25 m height). Point sources are introduced at an effective stack height which is the physical stack height plus plume rise. Plume rise is calculated according to Briggs (1971).

To avoid too sharp gradients resulting in numerical inaccuracies, it is required to horizontally spread out each source over a few neighbouring grid points.

Fig. II.2 shows a flow scheme of the KNMI-model.

For this specific study the following assumptions were made:

i. During night time and within a relatively short period of time, a stable layer develops, and extends up to a boundary-layer height, z_i. In the model, an average (but time dependent) over land value is used. In this layer air pollution is assumed to be transported by the (hourly) mean horizontal flow and by vertical diffusion only. At heights larger than z_i, the vertical diffusion is neglected so that pollutants are only horizontally displaced by the mean wind.

ii. When the atmospheric stability changes during the morning hours due to surface heating by the enhanced irradiation, the nocturnal inversion erodes and an unstable mixed layer develops. In the initial stage of the mixed-layer development vertical diffusion is still moderate. However, when the mixed-layer exceeds a fixed height (which is at present 600 m. The assumption is made that all sources emit below this value), it is assumed that turbulence mixes the pollutants throughout the layer in such short times that vertical concentration gradients will be considered negligible. During the inversion rise, the concentrations may change due to the entrainment of pollutants present in the adjacent upper layer (fumigation).

iii. In the late afternoon, a new ground-based stable layer develops. The pollution which - at that time - is present above 600 m will usually be advected outside the considered area during the following night so that it will not influence concentrations during the next day when a new mixed-layer develops. Sources present outside the area are not taken into account.

The vertical resolution is taken to be 50 m up to the maximum height of 600 m (12 layers).

The horizontal gridcells have an area of 20 km x 20 km. Because of the pseudospectral method used, the number of gridcells has to be 2^n, for this study resulting in 16 x 16 gridpoints (n = 4). Consequently, the KNMI-model has a smaller area in this study than the other models.

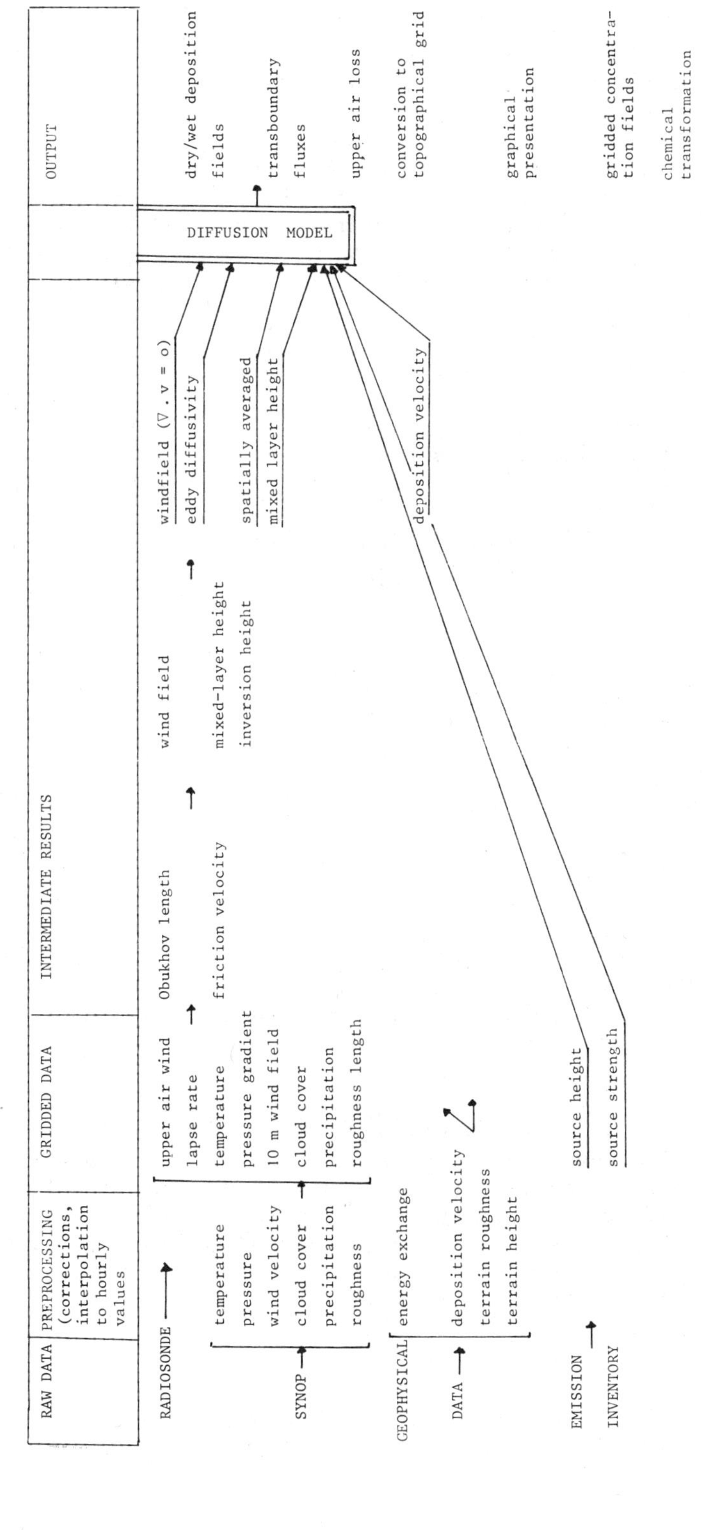

Fig. II.2 Flow scheme: KNMI-model.

Description of the windfield

The surface wind observations are interpolated by the use of the optimum interpolation method (Cats, 1980).

For the surface temperature, the optimum interpolation method is used too. The temperature field over land is determined from observations over land only (the higher situated stations are excluded). The air temperature over sea is derived from light-vessel (or platform) observations. As a consequence, the interpolation yields less accurate temperatures in coastal and elevated areas.

The basic components of the wind profile are:

- the surface-layer wind profile.
  In steady and homogeneous conditions the profile is given by the empirical relation

$$U_s(x,y,z) = U_c \frac{\ln(z/z_o)-\Psi_1(z/L)}{\ln(10/z_o)-\Psi_1(10/L)}, \qquad (II.3)$$

where Ψ_1 is given by

$$L < 0: \Psi_1 = 2\ln(\frac{1+x}{2}) + \ln(\frac{1+x^2}{2}) - 2\,\mathrm{arctg}(x) + \pi/2,$$

$$x = (1 - \frac{15z}{L})^{\frac{1}{4}} \qquad (II.4)$$

$$L > 0: \Psi_1 = -4.7\ z/L.$$

In II.3, U_s is chosen such that it passes through the (interpolated and local roughness length corrected) wind speed at 10 m, U_c. The dependence on the horizontal coordinates is through U_c, L and $z_o(x,y)$.

- The geostrophic wind profile.
  The spatial derivatives of the surface pressure field provide the surface level geostrophic wind U_{g0}. The 850 mbar wind velocity is linearly interpolated, and provides the 850 mbar geostrophic wind, U_{g1}. The geostrophic profile is obtained by linear interpolation of U_{g0} and U_{g1} according to

$$U_g(x,y,z) = \frac{(z_1-z)U_{g0} + (z-z_s)U_{g1}}{z_1 - z_s}, \qquad (II.5)$$

where z_1 and z_s are the levels where U_{g0} and U_{g1} are determined (respectively at 1.5 m and ~ 1500 m).

The two profiles, U_s and U_g are now integrated into one final profile by taking linear combinations,

$$U = \alpha_a U_s + \alpha_g U_g \tag{II.6}$$

The weighting functions α_s and α_g are determined according to a prescribed scheme.

In general the horizontal wind field thus obtained will not be divergence free, i.e.

$$\partial U/\partial x + \partial V/\partial y \neq 0. \tag{II.7}$$

On the one hand the imbalance is caused by measuring errors and model assumptions, and on the other by the presence of vertical movement which, according to the full non-divergence equation for an incompressible fluid, is given by:

$$W(z) = - \int_o^z \left(\frac{\partial U}{\partial x} + \frac{\partial V}{\partial y}\right) dz'. \tag{II.8}$$

It is expected that over more or less flat terrain the (hourly) mean vertical motion will be small, and will presumably not exceed the involved inaccuracies. Therefore the divergence which the derived U, V field might possess is removed by a procedure given by Endlich (1967). The use of divergence-free fields avoids numerical errors which arise when an accumulation of pollutant concentrations occurs in locally strong convergent areas. Steep gradients may then result herethrough which affect the numerical accuracy unnecessarily. For more information hereover the reader is referred to Van Dop et al (1982).

Vertical diffusivity and mixing height

The vertical diffusivity is determined by a method developed by Brost and Wyngaard (1978) using L and u_* derived from surface observations. The inversion height z_i is determined in the following manner. Both the nocturnal inversion height and the day-time mixed layer height will be denoted by z_i. A steady state expression is used for the determination of the nocturnal boundary-layer height (Nieuwstadt, 1981 a, b). It is based on the similarity expression:

$$z_i = 0.4(u_* L/f)^{\frac{1}{2}}, \tag{II.9}$$

(Brost and Wyngaard, 1978). It is modified such that in neutral conditions the expression for the neutral boundary-layer height results. Hence,

$$z = \frac{0.3\ u_*/f}{1 + 1.9\ z\ /L_i} \qquad \text{(II.10)}$$

After solving z_i form (II.10) we get:

$$z_i = 0.26\ L\ \{(1 + 2.04\ 10^4\ u_*/L)^{\frac{1}{2}} - 1\}. \qquad \text{(II.11)}$$

Although through the dependence on $u_*(t)$ and $L(t)_x$ the z_i-values vary from hour to hour, this variation is neglected in stable conditions and a time-averaged nocturnal boundary-layer height is used. Its spatial variation will be maintained however. It should be noted that (II.9) may also be used during daytime (daytime is defined as the period between one hour after sunrise and one hour before sunset), in stable, predominantly overcast conditions (cloud cover $N > 5$).

Stable situations during sunny conditions are assumed not to occur or, if reported, are to be ascribed to observational error. In unstable conditions ($L < 0$), it is assumed that a mixed layer exists, the height of which is non-descending. Buoyancy as well as mechanical forces will cause the inversion rise (Tennekes, 1973). We return here to the equations:

$$\partial\Theta/\partial t = (\overline{\Theta w}_o - \overline{\Theta w}_i)/z_i \qquad \text{(a)}$$

$$\Delta\ \partial z_i/\partial t = -\overline{\Theta w}_i, \qquad \text{(b)}$$

$$\partial\Delta/\partial t = \gamma\ \partial z_i/\partial t - \partial\Theta/\partial t, \qquad \text{(c)} \qquad \text{(II.12)}$$

$$-\overline{\Theta w}_1 = C_1\ \overline{\Theta w}_o - C_2(T/g)\ u_*^3/z_i, \qquad \text{(d)}$$

where C_1 and C_2 are constant (0.2 and 2.5 respectively), γ denotes the lapse rate of the layer beyond the inversion, Θ the mixed-layer potential temperature, and Δ the inversion strength. The heat fluxes at the surface and at the inversion height are denoted by Θw_o and Θw_i respectively. Once the initial conditions are given, the set equations II.12 can be solved with standard numerical methods provided that θw_o and γ are known functions of time.

We will use a somewhat simplified method where (hourly) stepwise increments in the boundary-layer height are calculated analytically. We consider now the two cases:

a. $C_1 \overline{\Theta w}_o >> C_2 (T/g) u_*^3 / z_i$.

This condition is equivalent with $h/\P L\P >> k\ C_2/C_1$ and is often satisfied in a daytime boundary layer with not too strong winds (Lamb, 1982). We thus neglect the second term in (II.12.d) and solve the set (II.12). A solution for Δ is:

$$\Delta = \frac{C_1}{1 + 2C_1} \gamma z_i + 0(z_i^{-6}), \tag{II.13.a}$$

where $0(..)$ means of the order of.
The last term in (II.13.a) is to be neglected and an analytical expression for z_i's derivative is then obtained:

$$\partial z_i/\partial t = \left(\frac{1 + 2C_1}{\gamma}\right) \cdot \frac{\overline{\Theta w}_o}{z_i}. \tag{II.13.b}$$

b. $C_1 \overline{\Theta w}_o << C_2 (T/g) u_*^3 / z_i$, or $0 < h/\P L\P << C_2/C_1$.

Here we have a situation where the mixed-layer growth is dominated by mechanical turbulence. The solution of (II.12) is obtained by neglecting the first term at the right hand side of Eq. (II.12.d):

$$\Delta = \tfrac{1}{2} \gamma z_i, \tag{II.14.a}$$

$$\partial z_i/\partial t = 2C_2/\gamma\ (T/g) u_*^3 / z_i^2, \tag{II.14.b}$$

We assume now that the general solution smoothly changes from (II.14) to (II.15) when $h/\P L\P$ varies from very large to small values. We therefore propose, as an approximate solution to (II.12), a linear combination of (II.13.b) and (II.14.b). The coefficients are chosen to be proportional to the magnitude of $h/\P L\P$.

A combination with the proper asymptotic behaviour is expressed by:

$$\partial z/\partial t = \frac{\overline{\Theta w}_o}{\gamma\, z_i (1 - (\frac{k\, C_2}{C_1})\frac{L}{z_i})} [1 + 2C_1 + 2k^2 C_2^2/C_1\ (L/z_i)^2]. \tag{II.15}$$

We obtain the inversion height at t + Δt from its previous value according to:

$$z_i(t + \Delta t) = z_i(t) + \Delta t\ \partial z_i/\partial t \tag{II.16}$$

where the time-step Δt is equal to 1 hr. Running along with (II.15) is the equation for rate of change of the boundary layer temperature,

$$\partial\Theta/\partial t = \frac{\overline{\Theta w}_o}{z_i} [1 + C_1 - k\ C_2\ L/z_i] \ , \tag{II.17}$$

which equation can be used to calculate the daily maximum temperature. For that purpose the boundary layer temperature is converted to the temperature at synoptic observation height (~1.50 m) by means of the relations:

$$\Theta - \Theta_o = 0.74 \frac{\overline{\Theta w}_o}{k\ u_*} \{\ln z/z_o - \Psi_2\ (z/L)\} \ , \tag{II.18}$$

where Ψ is given by (II.21.a). We have assumed that the mixed layer temperature is attained at a height of approximately 2.5 ¶L¶. The calculated temperatures can be compared with the observed temperatures, and may serve as a consistency check for the method so applied.

It should be noted that the methods used in this section are based on the dynamics of the atmospheric boundary-layer in horizontally homogeneous conditions. As a consequence large errors may occur in the boundary-layer height determination in the coastal zone for example.

Horizontal diffusion

The horizontal diffusion coefficient is set to zero.

Conversion

The chemical removal process is described by

$$S_c = -k_c \text{ conc } SO_2 \tag{II.19}$$

with $k_c = 0.01 \text{ hr}^{-1}$

Dry deposition

The deposition velocity V(z) is expressed by means of a resistance law,

$$V(z) = [r(z,1) + V_g^{-1}]^{-1}. \tag{II.20}$$

The resistance of the layer 1-z is expressed by r(z,1). The deposition velocity V_g is taken from literature (Sehmel, 1980). It depends both on the chemical properties of the surface and the pollutant. The chemical properties of the surface are more or less determined by its overgrowth. For that purpose a classification has been made of the various terrain types encountered in the region (Van Dop, 1983).

The resistance of the layer 1 - z, r(z,1), is given by:

$$r(z,1) = \frac{0.74}{k\,u_*} \{\ln(z) - \Psi_2(z/L) + \Psi_2(1/L)\}. \tag{II.21}$$

The stability function Ψ_2 is given by:

$$L < 0 : \Psi_2 = 2 \ln(\frac{1+y}{2}),\ y = (1 - 9\,\frac{z}{L})^{\frac{1}{2}} \tag{II.21.a}$$

$$L > 0 : \Psi_2 = -6.4\ z/L.$$

Wet deposition

The wet deposition is expressed by:

$$D = -k_w \text{ conc } SO_2 \tag{II.22}$$

with $k_w = a\,P$

where P is the precipitation rate and a is 0.3 mm^{-1}.

II.c The TNO(SAI)-model

The TNO (= Institute of Applied Scientific Research)/SAI (= Systems Applications Inc.) -model was originally developed by SAI, and is described by Reynolds et al (1979).

Basic features

The model is an Eulerian gridmodel with the main emphasis on calculating non-linear chemical processes such as those occurring during photochemical episodes. The basic formula solved is:

$$\frac{\partial c}{\partial t} + u\frac{\partial c}{\partial x} + v\frac{\partial c}{\partial y} + w\frac{\partial c}{\partial z} = \frac{\partial}{\partial y}\left[k_y \frac{\partial c}{\partial y}\right] + \frac{\partial}{\partial z}\left[k_z \frac{\partial}{\partial z}\right] + S$$

The typical size of the area normally used for the SAI-airshed model is about 150 x 150 km^2, with grids of 2 x 2 km^2 to 10 x 10 km^2 and with several levels in the vertical direction. The total simulation is based on a period of one or two days. The model has been used for several regions in the United States and Europe, especially for regulation questions such as the determination of the most suitable place for a large power station, the impact made by certain car-exhaust regulations and the air-quality changes by different economical and industrial development scenarios.

For the numerical scheme, the method of fractional steps is employed. The advective part is solved using the SHASTA method (Boris and Book, 1973); for the diffusive part a standard finite difference technique is used and which is explicit for the horizontal diffusion and implicit for the vertical diffusion. The chemical reactions are treated according to a Newton iterative procedure.

Ground-level and point-source emissions are also considered. These emissions are instantaneously mixed in the grid into which they are emitted. For the point-source emissions the plume-rise formula of Briggs is used.

The model is divided into sub-models for each part of the calculation; for example a windfield sub-model, a chemical sub-model, a boundary value sub-model etc. The output of each sub-model can be considered separately before these outputs are used as input to the simulation program itself.

As its heart, the SAI Airshed Model System contains the Airshed Simulation Program, the input to which consists of 10 to 14 data files, depending on the program options chosen. Each file is created by a separate data preparation program. The output from the Airshed Simulation Program consists of three data results files and an execution trace report. The output data files can be further processed by any display and analysis programs and can also be used for restarting the simulation. Figure II.3 shows the Airshed Simulation Program and identifies its input and output files.

The SAI Airshed Model System can be described according to the five major tasks or functions that the user and the programs must perform:

- M.E.B. (meteorology, emissions, and initial and boundary conditions) file preparation

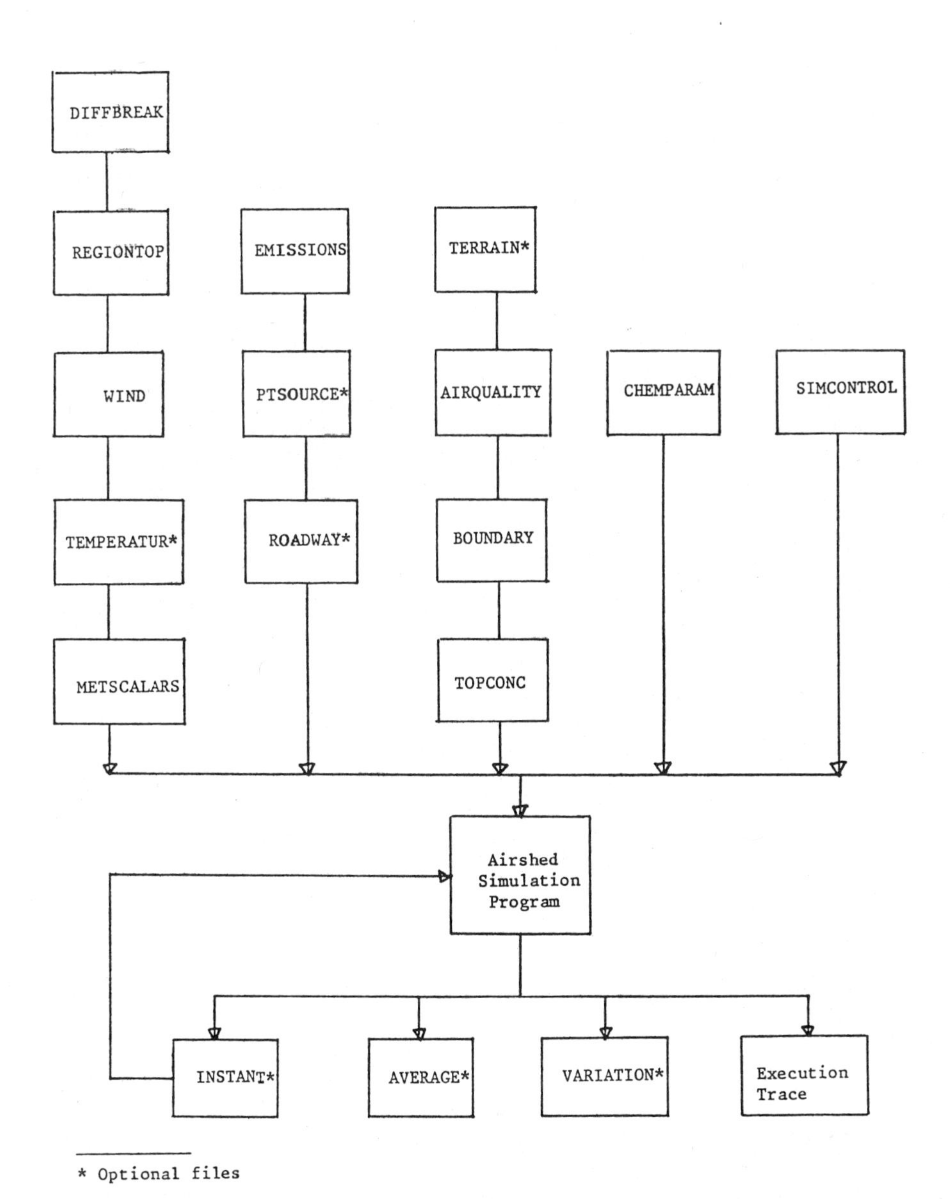

* Optional files

Fig. II.3 Airshed simulation program with input and output files.

- M.E.B. file segmentation
- Control data preparation
- Airshed simulation
- Display and analysis

Each of the five major functions is represented by a rectangular box in Figure II.4 which illustrates the flow of information through the system and shows the position of each major function within that system. The following sections of this chapter briefly discuss each function.

For the specific study described here, the airshed model has been modified slightly. The non-linear chemical reaction scheme has been replaced by a one specific-SO_2-linear reaction scheme to be comparable to the other models used in this study which cannot handle non-linear mechanisms. Although the process of dry deposition has been incorporated in the airshed-model, a special subroutine has been put into the model to keep track of the amounts deposited at the surface. An algorithm to handle wet deposition has also been incorporated. The horizontal dimensions of the model region were as follows: latitude 49°-54° N, longitude 3°-8° E. The top of the model region was 1650 m above the surface for the 1979 episode and 500 m for the 1980 episode. For both episodes, the horizontal grid dimensions were 10 km. For vertical resolution five layers of variable thickness were defined. The lowest layer (layer 1) was of constant thickness (50 m); layer 2, 3 and 4 were of equal thickness, such that the top of layer 4 coincided with the mixing height. The minimum thickness of layer 2, 3 and 4 however, was set at 150 m (1979 episode) and 50 m (1980 episode) respectively. Therefore, at night the mixing height was lower than the top of layer 4. This provision enabled the model to retain sufficient vertical resolution at the heights where most chimneys effectively emit. The model region was defined such that its bottom was coincident with the height of the real surface of the earth.

Description of the windfield

The wind field was constructed using the following interpolation procedure.

a. Surface wind data for every hour were horizontally interpolated using 1/r - weighting. The weighting factor was set to zero for $r > 75$ km and to $(25\ km)^{-1}$ for $r < 25$ km. This field is called the surface field SF.
b. From vertical soundings (typically every 6 or 12 hours), vertical wind profiles were constructed using a spline interpolation.

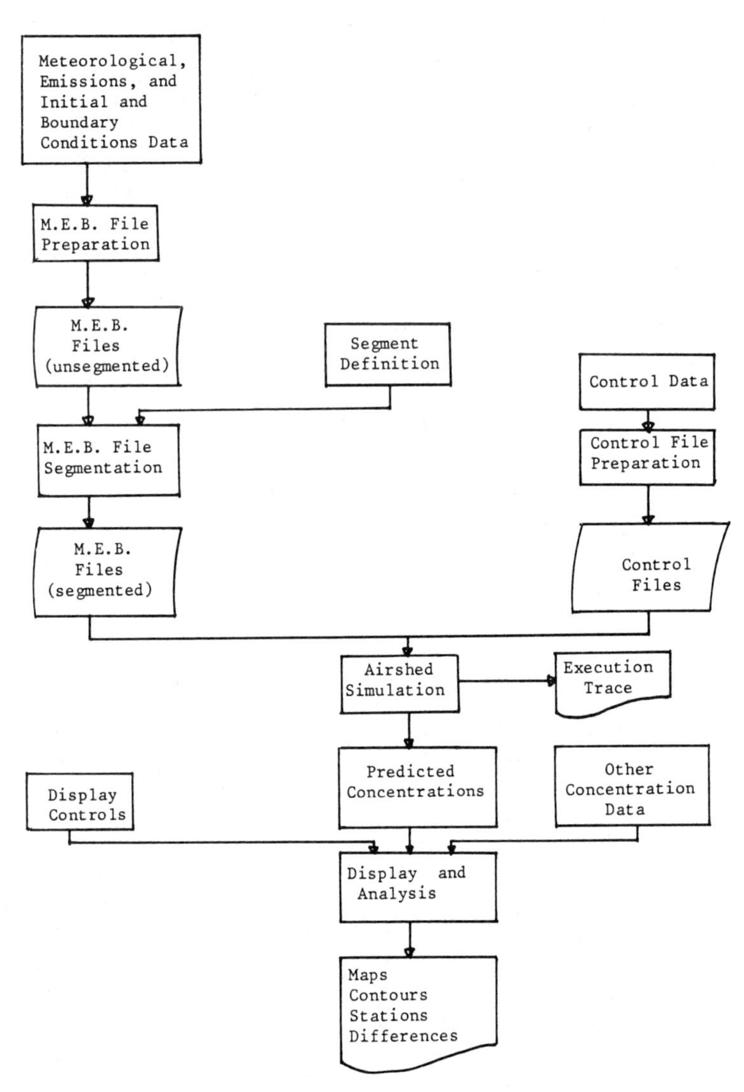

Fig. II.4 Information flow in the SAI Airshed Model System.

c. For sounding stations where measurements of vertical profiles were temporarily unavailable, a linear interpolation-in-time was performed provided that the time between the previous and later sounding was not more than 24 hours.
The set of vertical profiles thus obtained was not always sufficient for the interpolation procedure: an additional profile in a corner of the model region was supplemented when the procedure would otherwise have to extrapolate from distant soundings. This profile was constructed by subjective extrapolation of the available measured wind profiles.
d. For each layer (including the surface layer) the wind components (averaged over the layer height) were interpolated horizontally using 1/r - weighting. The weighting factor was set to zero for $r > 300$ km and to $(100 \text{ km})^{-1}$ for $r < 100$ km. This field is called the profile field P_i, where i refers to the layer number.
e. The 3-dimensional wind field is then constructed by linear combination of the surface and profile fields:

$$W_i = (SF + (P_i - P_1) \cdot F(z_i) + P\ (1 - F(z_i))$$

$$F(z_i) = 1 - \frac{z-50}{G-50} \qquad \text{for } 50 \text{ m} < z < G$$

$$F(z_i) = 0 \qquad \text{for } z \geqq G$$

G = 300 m for mixing height < 300 m
G = mixing height for 300 ≦ mixing height ≦ 1000 m
G = 1000 m for mixing height > 1000 m

Expressed literally, this means:
At the surface, the ground field is taken. The profile field is taken above height G, where G equals the mixing height, with a minimum of 300 m and a maximum of 1000 m. Between the surface and G, a linear combination is taken of the profile field and the profile field that has been adjusted using the local surface wind. It was verified that the vertical wind speed, calculated from the horizontal divergence, did not give rise to important systematic vertical motions. The vertical wind velocity was finally set to zero. In this way, the vertical exchange of matter between layers arising from local divergences was suppressed.

Vertical diffusion and mixing height

For the calculation of vertical diffusivities, the Monin-Obukhov length and the friction velocity were estimated from the surface wind speed, cloudiness and surface roughness. From these parameters the vertical diffusivity profile was calculated according to Businger and Arya (1974) for stable conditions. For neutral and unstable conditions, the results of Deardorff (1972) were used.

Above the mixing height, K_z was 0.01 m^2/s, and the mixing height was estimated by evaluation of vertical soundings and surface data. For the 1979 and 1980 episodes, the mixing height was taken to be constant in space. The mixing height at noon was directly estimated from the vertical soundings. For the 1980 episode, the vertical profiles were ambiguous in this respect so that a rather subjective choice had to be made. The night time mixing height was estimated from the average geostropical wind speed according to $h_n = v_g^2 + 49$. Fig. II.5 and II.6 exhibit the mixing heights used for the 1979 and 1980 episodes, including the ones of RIV and KNMI. Figure II.7 shows the mixing heights used for the 1981 episode for RIV and KNMI.

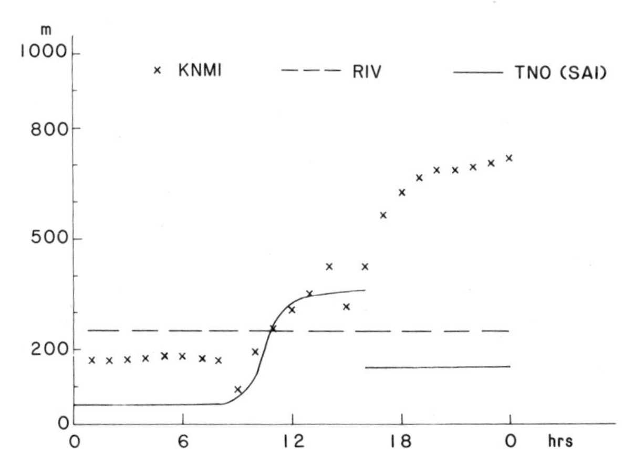

Fig. II.5 Mixing height for episode α (22-2-1979).

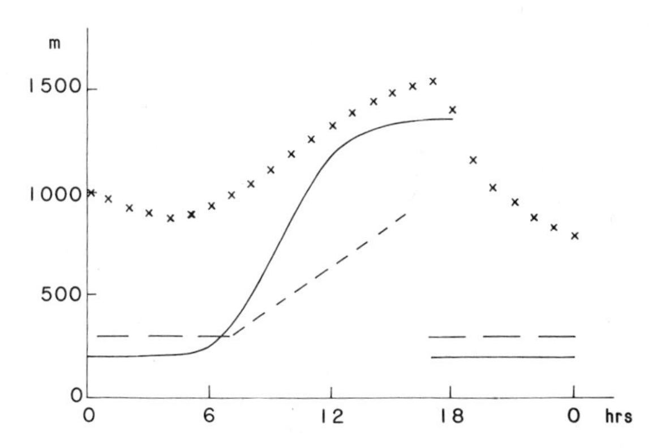

Fig. II.6 Mixing height for episode β (14-5-1980).

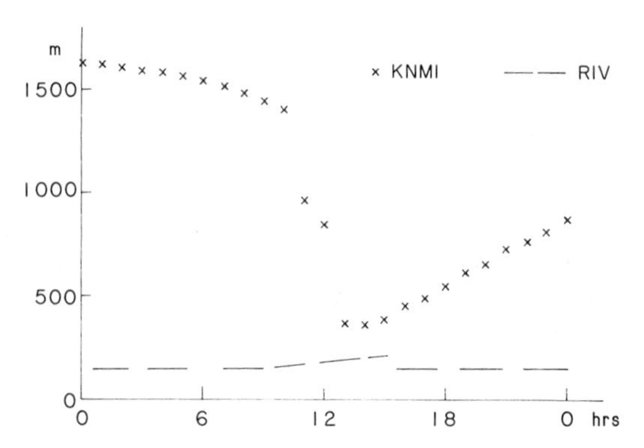

Fig. II.7 Mixing height for episode γ (15-12-1981)

The values of the mixing height shown in Figs. II.5 - II.7 are in principle based on the same meteorological data base with the exception of the RIV-values (see page 265). The differences, which will have a substantial effect on calculated concentrations are exclusively caused by different assumptions made to determine the mixing height. It is observed that these assumptions are all physically reliable and that criteria for evaluation were not available.

Horizontal diffusion

For this study the horizontal diffusion has been set to a constant value of $K_H = 50\ m^2/s$.

Conversion rate

The conversion of SO_2 into sulphate was described as

$$\frac{d\ SO_2}{dt} = -\ (k_1 + k_2.R).SO_2$$

with
$k_1 = 8.3 \times 10^{-5}\ min^{-1}$
$k_2 = 8.3 \times 10^{-4}\ min^{-1}$
R is the photolysis rate constant for NO_2.

So the conversion rate changes with time. It is approximately $1\%\ hr^{-1}$.

Dry deposition

Dry deposition is calculated using a dry deposition velocity parameter V_d:

$$V_d = \frac{1}{R_t + R_{si}}$$

with $R_t = u_r u_*^{-2} + B^{-1} u_*^{-1}$

$$B^{-1} = 2.2\, u_*^{-1/3}$$

To determine the surface resistance for grid square j, R_{sij}, we calculate the average surface uptake velocity $V_{sij} = (R_{sij}^{-1})$ from:

$$V_{sij} = \Sigma\, \alpha_{jk}\, \beta_k\, V_{si}^r$$

where
αjk = fraction of grid square j represented by land use k,
β_k = a factor that adjusts the reference surface uptake velocity to that for land use k,
V_{si}^r = reference surface surface uptake velocity for pollutant i.

Values of β_k for a central business district, suburban residential, and rural/agricultural land use categories have been estimated at 0.2, 0.5 and 1.0, respectively (Killus et al. 1977). Because the original value of V_{si}^r seemed rather low for SO_2, it was increased by a factor 3 to 0.034 m/s.

Wet deposition

The wet deposition is described according to

$$\left(\frac{dSO_2}{dt}\right)_w = -\lambda\, P\, SO_2$$

with $\lambda = 0.22\ \text{mm}^{-1}$
and P is the precipitation rate (mm/hr).

II.d The RIV (National Institute of Public Health) - model

The *original* RIV model has been described in detail by Van Egmond and Kesseboom (1983). *For the purpose of the intercomparison the original model had to be modified in order to comply as far as possible with the standardized input data.*

The Lagrangian PUFF-model is based on the transport of a large number of (two layer) Gaussian puffs:

$$C(x,y) = f \frac{M}{2\pi \sigma_y^2 h} \exp \left[\frac{-(x^2 + y^2)}{2\sigma_y^2}\right]$$

in which the symbols have their usual meaning and f is a factor which describes the concentration distribution in z-direction. At smaller distances the value of f is given by the conventional Gaussian plume model accounting for effective plume height and reflections at ground- and inversion level. For the plume rise a modified Briggs scheme is used.

The required values for σ_z are obtained from the conventional empirical Pasquill functions where the Pasquill stability class is derived from Obukhov length L according to the scheme given by Golder (1972). Vertical stratification is given by three layers,

- mixing layer, assumed to be constant during the night and increasing during daytime with a speed given by the acoustic sounder measurement,
- reservoir layer in which the pollution of higher sources is emitted, especially during the night,
- surface layer of 25 m in which the vertical turbulent diffusion is restricted by interactions with the earth surface, resulting in a limited mass transfer from the mixing layer to the ground.

The mixing- and reservoir layer masses are treated numerically, the surface layer concentration gradient, as represented by f is derived analytically from aerodynamic- and surface deposition resistances (Wesely and Hicks, 1977). In the original model these resistances are obtained from L and u_* which in turn are derived from solar radiation and 10 m windspeed (z_o = 0.05 cm). During the night the empirical relation $L = 1100\ u_*^2$ (Venkatram, 1980) is used.

The puffs are generated with initial σ_{yo} such that the source diameter is $4\sigma_{yo}$. The puffs are released with inter-puff distances of 10 km. At the final projection of the individual puffs to the Eulerian mapping grid, additional puffs are generated in the direction of the wind vector which is assigned to every puff at the time and location of emission. As such, the puffs actually represent plume segments over 10 km intervals and together for a continuous plume. The plume segment approach enlarges the number of parameters which has to be handled for every puff, but on the other side reduces the number of required puffs and thus reduces computation time. Finally the following 10 parameters are assigned to every puff:

- x-position
- y-position
- σ_y-standard deviation of puff
- t^y total time of travel
- f^p fraction of concentration affecting ground level
- mass in reservoir layer
- mass in mixing layer
- x-component plume-segment vector (wind at time of emission)
- y-component plume-segment vector
- effective height H of initial emission

The PUFF-model was designed to describe mesoscale air pollution transport on the basis of routinely (real time) available input data:

- wind direction and - speed at 10 m level
- wind direction and - speed at levels between 150 m - 300 m
- standard deviation of wind at various levels
- (solar) radiation
- mixing height (acoustic sounder)

In order to restrict the input data to standard synoptical meteorological data, the model was essentially varied, resulting in suboptimum model performance. In contrast to the original model, stability classes according to Pasquill were derived from windspeed and cloud cover. The flow diagram for the modified model is given in Fig. II.8. The actual values for the parameters used in the RIV-model in the three case studies are given in Tables II.2, II.3 and II.4.

Windfield description

The wind direction is inter- or extrapolated by fitting the Ekman-Taylor spiral wind profile.

It is calculated at heights representative of the mixing layer and reservoir layer. Horizontal interpolations are made by means of a negative exponential weighting scheme. Horizontal divergence is reduced by application of the iterative scheme of Endlich (1967). As opposed to the original version of the model, the following steps were carried out:

i. From Pasquill-stability classes the exponent m in the power law vertical windprofile is derived, according to the advised values in the Netherlands for Gaussian model applications. Herein m = 0.10 at class A, m = 0.16 at class D and m = 0.30 at stability class F.

ii. Standard deviation of wind direction, which is used in the original PUFF-model for modelling puff-growth, is also derived from stability class, according to H. Gland (1982).

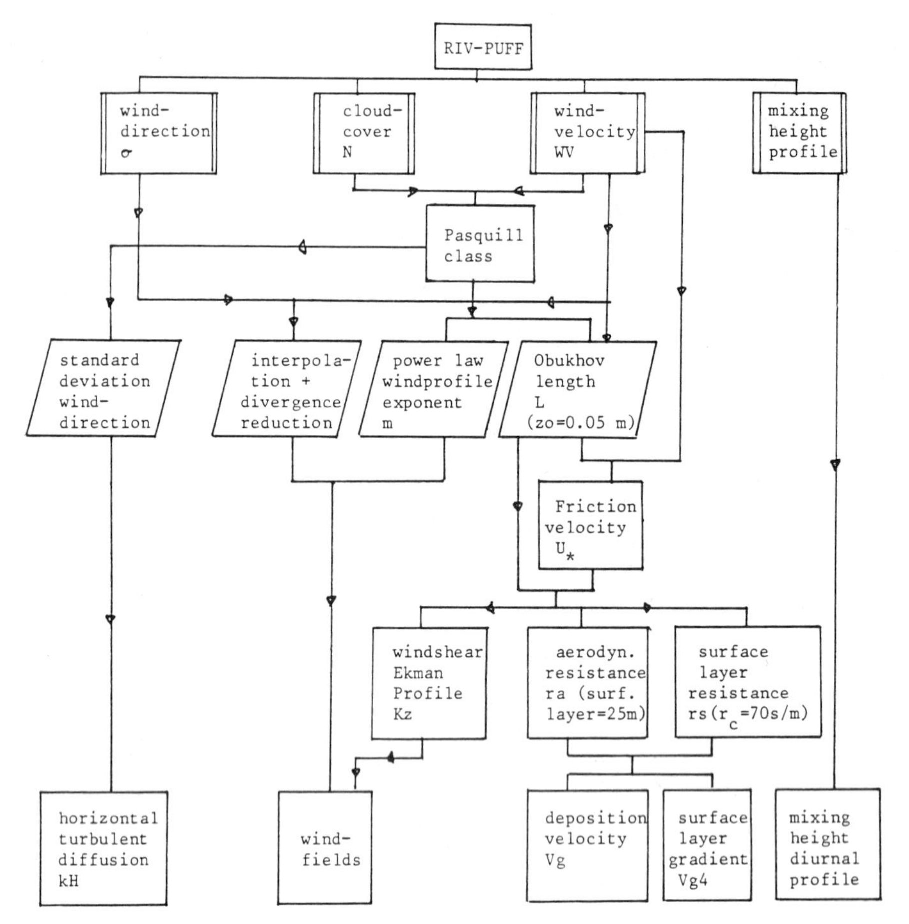

Fig. II.8 Flow scheme: RIV-model.

Vertical diffusion and mixing height

The vertical diffusion was calculated differently from the original model:

According to Golder (1972) Pasquill-stability was converted to Obukhov-length L. From wind velocity and Obukhov-length friction velocity u_* is derived iteratively. From L and u_*, K_z at z = 25 m was computed. Based on the results of earlier experiments K_z was taken as representative for K_{zm}, the "bulk-turbulent diffusity", which, as an effective average, is used for estimation of wind shear, according to the Ekman-Taylor profile. In this way the spiral profile for vertical interpolation of winddirection as applied in the original model can be used.

Table II.2 RIV parameter - values for episode α (1979) for De Bilt, The Netherlands.

Time	$\bar{N}$ (1/8)	$\bar{WV}$ knots	Pasq	L m	U_* m/s	K_z m^2/s	SDwind degrees	m	V_D cm/s	V_D4	H m
1	6	2	F	10	0.08	0.2	3	0.3	0.14	0.48	250
2	5	2	F	10	0.08	0.2	3	0.3	0.14	0.48	250
3	5	3	F	10	0.08	0.2	3	0.3	0.14	0.48	250
4	7	3	D	10	0.08	0.2	10	0.16	0.14	0.48	250
5	8	3	D	10	0.08	0.2	10	0.16	0.14	0.48	250
6	8	3	D	10	0.08	0.2	10	0.16	0.14	0.48	250
7	8	3	D	10	0.08	0.2	10	0.16	0.14	0.48	250
8	8	3	D	10	0.08	0.2	10	0.16	0.14	0.48	250
9	8	3	D	10	0.08	0.2	10	0.16	0.14	0.48	250
10	8	3	D	-160	0.13	6.0	10	0.16	0.44	0.89	250
11	8	3	D	-160	0.13	6.0	10	0.16	0.44	0.90	250
12	8	5	D	-160	0.17	7.8	10	0.16	0.53	0.91	250
13	8	5	D	-140	0.17	8.3	10	0.16	0.53	0.91	250
14	8	6	D	-140	0.21	10.2	10	0.16	0.60	0.89	250
15	8	6	D	-140	0.21	10.2	10	0.16	0.60	0.89	250
16	8	6	D	-280	0.20	7.6	10	0.16	0.57	0.89	250
17	8	6	D	-280	0.20	7.6	10	0.16	0.57	0.76	250
18	8	6	D	40	0.20	0.4	10	0.16	0.50	0.76	250
19	8	7	D	30	0.23	0.4	10	0.16	0.40	0.76	250
20	8	8	D	40	0.23	0.6	10	0.16	0.50	0.76	250
21	8	8	D	40	0.23	0.6	10	0.16	0.50	0.76	250
22	8	8	D	40	0.23	0.6	10	0.16	0.50	0.76	250
23	8	8	D	40	0.23	0.6	10	0.16	0.50	0.76	250
24	8	8	D	40	0.23	0.6	10	0.16	0.50	0.76	250

Mixing heights could not be derived from the standard meteorological data. Consequently estimates for the mixing height profile were based on acoustic sounder measurements.

The horizontal diffusion

The horizontal distribution is given by σ_y and the increase σ_y is found as:

$$\sigma^2_{y\ t+\Delta t} = \sigma^2_{y\ t} + 2\ K_H\ \Delta t$$

Table II.3 RIV parameter - values for episode β (1980).

Time	$\bar{N}$ (1/8)	$\bar{WV}$ knots	Pasq	L m	U_* m/s	K_z m^2/s	SDwind degrees	m	V_D cm/s	V_D4	H m
1	0	13	D	180	0.40	3.4	10	0.16	0.77	0.88	300
2	0	13	D	180	0.40	3.4	10	0.16	0.77	0.88	300
3	0	12	D	150	0.37	2.7	10	0.16	0.74	0.87	300
4	0	12	D	140	0.37	2.7	10	0.16	0.74	0.87	300
5	1	12	D	140	0.37	2.7	10	0.16	0.74	0.87	300
6	1	12	D	140	0.37	2.7	10	0.16	0.74	0.87	300
7	0	12	C	-300	0.41	15.3	15	0.16	0.82	0.92	300
8	0	13	C	-300	0.44	16.5	15	0.16	0.85	0.93	370
9	1	14	C	-280	0.47	17.9	15	0.16	0.87	0.93	440
10	1	14	C	-200	0.48	20.5	15	0.16	0.88	0.94	510
11	1	15	C	-160	0.48	22.2	15	0.16	0.89	0.94	580
12	1	15	C	-140	0.50	24.3	15	0.16	0.90	0.94	650
13	1	15	C	-140	0.50	24.3	15	0.16	0.90	0.94	720
14	1	16	C	-160	0.55	25.4	15	0.16	0.93	0.94	790
15	1	15	C	-180	0.55	24.3	15	0.16	0.93	0.94	860
16	2	16	C	-200	0.55	23.4	15	0.16	0.93	0.94	930
17	2	16	C	-200	0.55	23.4	15	0.16	0.93	0.94	300
18	2	15	C	-260	0.50	19.5	15	0.16	0.89	0.93	300
19	2	15	C	-260	0.50	19.5	15	0.16	0.89	0.93	300
20	1	15	D	-300	0.50	18.7	10	0.16	0.89	0.93	300
21	1	13	D	160	0.41	3.2	10	0.16	0.78	0.88	300
22	0	12	D	140	0.37	2.7	10	0.16	0.74	0.87	300
23	0	11	D	120	0.34	2.2	10	0.16	0.70	0.86	300
24	0	11	D	120	0.34	2.2	10	0.16	0.70	0.86	300

K_H is the time dependent apparent diffusivity and is computed for every puff-travel time t from the Lagrangian correlation function $R_1(\xi) = e^{-\xi/t_L}$ and the measured cross wind turbulence v_m^2.

The conversion, dry deposition and wet deposition

One percent sulphur dioxide is removed per hour in order to simulate the influence of the conversion from SO_2 to SO_4. The dry deposition is a function of atmospheric stability and surface resistance. The wet deposition process has not been considered.

Table II.4 RIV parameter - values for episode γ (1981).

Time	$\bar{N}$ (1/8)	$\bar{WV}$ knots	Pasq	L m	U_* m/s	K_z m^2/s	SDwind degrees	m	V_D cm/s	V_D4	H m
1	7	10	D	100	0.30	1.7	10	0.16	0.65	0.84	150
2	7	10	D	100	0.30	1.7	10	0.16	0.65	0.84	150
3	7	9	D	80	0.26	1.2	10	0.16	0.59	0.82	150
4	7	10	D	100	0.30	1.7	10	0.16	0.65	0.84	150
5	7	10	D	100	0.30	1.7	10	0.16	0.65	0.84	150
6	7	9	D	80	0.26	1.2	10	0.16	0.59	0.82	150
7	7	9	D	80	0.26	1.2	10	0.16	0.59	0.82	150
8	6	9	D	80	0.26	1.2	10	0.16	0.59	0.82	150
9	6	8	D	40	0.23	0.6	10	0.16	0.50	0.76	150
10	6	7	C	-220	0.23	9.5	15	0.16	0.62	0.90	160
11	5	7	C	-140	0.25	12.1	15	0.16	0.66	0.92	170
12	5	6	C	-60	0.22	15.2	15	0.16	0.64	0.93	180
13	5	6	C	-60	0.22	15.2	15	0.16	0.64	0.93	190
14	5	5	C	-40	0.23	19.0	15	0.16	0.67	0.94	200
15	6	5	C	-140	0.21	10.2	15	0.16	0.60	0.91	210
16	6	5	C	-140	0.21	10.2	15	0.16	0.60	0.91	150
17	6	6	D	20	0.14	0.2	3	0.30	0.29	0.63	150
18	6	6	E	20	0.14	0.2	3	0.30	0.29	0.63	150
19	6	7	D	40	0.19	0.5	10	0.16	0.44	0.79	150
20	6	8	D	40	0.23	0.6	10	0.16	0.50	0.76	150
21	5	9	D	60	0.26	1.0	10	0.16	0.57	0.80	150
22	5	10	D	100	0.30	1.7	10	0.16	0.65	0.84	150
23	4	10	D	100	0.30	1.7	10	0.16	0.65	0.84	150
24	4	10	D	100	0.30	1.7	10	0.16	0.65	0.84	150

III. PHASE I - DISPERSION FROM A HYPOTHETICAL SOURCE IN A HOMOGENEOUS FLOW FIELD

a. The model experiments of phase I

The aim of these model experiments was to test the basic algorithm of the different models for simple problems and to discuss the behaviour of each model in comparison with the corresponding analytical solution, for test cases A and B, or with the other models, for test case C.

Table III.1 Numerical values in the three test cases.

	case		A	B	C
u	wind speed	ms^{-1}	10	1	10
K_z	vertical eddy diff.	m^2s^{-1}	5	0.1	H/100 for z < H, for z > H, K_z = 0
H	inversion height	m	∞	100	H = 100 (t+1)
Q	source strength	kgs^{-1}	1	1	1
h	source height	m	150	0	300

The test cases A and B deal with a single point source and uniform and steady conditions and hence primarily test vertical diffusion. In test case A dispersion in a strong, uniform windfield and turbulence was considered. The along-wind turbulent diffusion was neglected in all Eulerian models.

In test case B a dispersion process in low wind, uniform conditions was simulated. The stationary solution of these two cases can be compared with an analytical Gauss-solution (see paragraph b).

In test case C an unsteady situation was considered where the vertical diffusivity profile is time-dependent in order to model an early morning inversion rise (fumigation). Chemical and deposition processes were not simulated in phase I. Table III.1 shows the numerical values which were used in the three test cases.

The area for test cases A and C extends from the surface to 1000 m height with a vertical resolution of 100 m. The horizontal extension is roughly 200 km with a resolution of 10 km. In test case B the vertical extension is different from the other two cases. The upper boundary of the area is situated at a height of 100 m. The vertical resolution is 10 m. The inversion layer at the upper boundary reflects the diffusion material totally.

b. The analytical model

The diffusion equation III.1

$$\frac{\partial C}{\partial t} + \frac{\partial}{\partial x_i}(v_i C) = \frac{\partial}{\partial x_i}\left(K_{x_i}\frac{\partial C}{\partial x_i}\right) + S \qquad \text{(III.1)}$$

can be solved assuming that:

i) $\frac{\partial C}{\partial t} = 0$ — stationarity

ii) $v = w = 0; \frac{\partial u}{\partial x} = \frac{\partial u}{\partial y} = \frac{\partial u}{\partial z} = 0$ — homogeneous windfield

iii) $\frac{\partial}{\partial x_i} (K_{x_i}) = 0$ — homogeneous eddy diffusivity

iv) $K_x \frac{\partial^2 C}{\partial x^2} << u \frac{\partial C}{\partial x}$ — diffusion in x-direction ignored

v) $S = 0$ — no sources and sinks in the dispersion area.

In addition hereto, the model calculation includes the assumption that the distribution of diffusing material in the plume is Gaussian in the crosswind and vertical directions. The standard deviations of these distributions are functions of source-distance and atmospheric stability.

This leads to the following equation:

$$U \frac{\partial C}{\partial x} = K_y \frac{\partial^2 C}{\partial y^2} + K_z \frac{\partial^2 C}{\partial z^2} \tag{III.2}$$

In connection with the boundary conditions

a) single point source at $x = y = 0$ and $z = h_s$; $C = \frac{Q}{v} \cdot \delta\,(0,0,h_s)$

b) disappearing concentrations for large x,y; $x,\ y \to \pm\infty \to C \to 0$ (III.3)

and

c) reflecting top and/or ground; $\left. \begin{array}{l} z = 0 \\ z = H \end{array} \right\} \to K_z \frac{\partial C}{\partial z} = 0$

one gets the following solution III.4:

$$C(x,y,z) = \frac{Q}{U} \frac{1}{2\Pi\,\sigma_y} \exp\left(-\frac{y^2}{2\sigma_y^2}\right) F\,(x,z) \tag{III.4}$$

where

$$F(x,z) = \frac{1}{2\Pi\,\sigma_z}\left\{\exp\left(-\frac{(z-h_s)^2}{2\,\sigma_z^2}\right) + \exp\left(-\frac{(z-h_s)^2}{2\,\sigma_z^2}\right)\right\} \qquad \text{(III.5)}$$

for an open upper boundary or

$$F(x,z) = \frac{1}{2\Pi\,\sigma_z}\sum_{j=-\infty}^{+\infty}\left\{\exp\left(-\frac{(z-h_s-2jH)^2}{2\,\sigma_z^2}\right) + \exp\left(-\frac{(z+h_s-2jH)^2}{2\,\sigma_z^2}\right)\right\} \qquad \text{(III.6)}$$

for a reflecting lid a height H.

The standard deviations σ_y and σ_z are defined as

$$\sigma_y = \sqrt{2\,K_y\,\frac{x}{u}} \qquad \text{(III.7)}$$

$$\sigma_z = \sqrt{2\,K_z\,\frac{x}{u}}$$

C(x,y,z) describes the concentration at each point of the dispersion area. In order to compare this solution with the results of a box-model one has to calculate a mean concentration for each box. These values were obtained by numerical integration.

c. Differences between the models

In order to be able to discuss the results of the calculations of phase I, it appears necessary to point out some of the characteristics of the four models. There have been some major differences between the models in dealing with the given problems:

1. Each model treats the horizontal diffusion differently. In fact, for this test case as input a value of $K_y = 10\ m^2/s$, had been given. This value, however is very small compared to the gridsize used. In the case of the TNO (SAI)-model this led to a strong influence of the horizontal diffusion which had to be reduced. Accordingly a gridsize of $\Delta x = 10$ km and $\Delta y = 2$ km was used. Then the concentrations were integrated in order to get a 10 x 10 km^2 box-average concentration. Also the THD-model gives concentrations averaged over a box of 10 x 10 km^2.

2. The KNMI-model neglects lateral diffusivity ($K_y = 0$). Therefore the results were compared with crosswind integrated analytical results. These can easily be compared with the 3-dimensional case, because nearly all the released material in each of the test cases stays within the centre plane gridcells. With the relation

$$C_n\ (x,0,z) = C_n^y\ (x,z)/\Delta y \tag{III.8}$$

the 2-dimensional results $C_n^y\ (x,z)$ can be transformed into the 3-dimensional results $C_n\ (x,0,z)$ where Δy is the width of the gridcell. (At the KNMI a 2-dimensional version of the original model was developed and used for phase I. Because there is no horizontal diffusion in any of the two models, this does not create any problems with the intercomparison.) Another characteristic of the KNMI-model is the fact that the source is spread over three gridpoints with the major contribution by the centre one.

3. The RIV-PUFF-model produces grid-points solutions from which box-averages concentrations were obtained by numerical integration over three gridpoints.

d. Discussion of the results of phase I

The comparison has been carried out for the vertical cross-section through the plume centre.

Case A

The tables III.2, III.3, III.4 and III.5 show the results (in $\mu g/m^3$) of the four models plotted above the values of the analytical solution. Furthermore tables III.6 and III.7 give information on the correlation coefficients (linear and logarithmic), the root mean square error as defined by

$$\sqrt[2]{\frac{1}{N}\sum_{i=1}^{N} (C_{1i} - C_{2i})^2}$$

the vertically integrated concentration etc. The results of the model calculations indicate that all models lose substance in the plume centre depending on the distance from the source. The analytical solution show also a slight decrease with distance. This is very probably due to truncation errors. In addition some material is transported into the y-dimension.

The agreement between the results of the THD-model and the analytical solution is in general very good. Although it is also good for the KNMI-model, the vertical integrated concen-

(text continued on page 194)

Table III.2 THD-results case A.

Concentrations in $\mu g/m^3$

Z in m \ X in km	5	15	25	35	45	55	65	75	85	95	105	115	125	135	145	155	165	175	185	195
950	0,00 0,00	0,01 0,00	0,02 0,00	0,05 0,00	0,09 0,02	0,15 0,06	0,21 0,12	0,29 0,22	0,38 0,34	0,47 0,48	0,57 0,64	0,66 0,80	0,76 0,98	0,85 1,15	0,93 1,33	1,01 1,50	1,09 1,68	1,16 1,85	1,22 2,01	1,28 2,17
850	0,01 0,00	0,03 0,00	0,07 0,00	0,15 0,02	0,26 0,09	0,40 0,21	0,56 0,39	0,74 0,59	0,93 0,82	1,13 1,07	1,32 1,32	1,52 1,57	1,70 1,82	1,87 2,05	2,04 2,28	2,19 2,50	2,33 2,71	2,46 2,91	2,58 3,10	2,68 3,28
750	0,02 0,00	0,09 0,00	0,22 0,03	0,41 0,14	0,65 0,37	0,93 0,69	1,24 1,04	1,55 1,42	1,87 1,79	2,18 2,16	2,47 2,50	2,75 2,83	3,01 3,14	3,25 3,42	3,17 3,69	3,68 3,93	3,86 4,16	4,02 4,37	1,17 4,56	1,29 4,73
650	0,08 0,00	0,29 0,02	0,63 0,21	1,05 0,65	1,53 1,24	2,03 1,85	2,52 2,45	2,99 3,00	3,11 3,50	3,84 3,95	4,21 4,36	4,55 4,72	4,85 5,05	5,12 5,34	5,35 5,59	5,56 5,82	5,75 6,03	5,91 6,21	6,05 6,37	6,17 6,52
550	0,30 0,00	0,91 0,24	1,70 1,13	2,55 2,28	3,37 3,33	4,12 4,23	4,79 4,98	5,38 5,61	5,89 6,14	6,32 6,59	6,70 6,96	7,01 7,28	7,29 7,56	7,52 7,78	7,71 7,98	7,87 8,14	8,01 8,28	8,12 8,39	8,21 8,49	8,28 8,57
450	1,13 0,12	2,75 1,86	4,35 4,34	5,73 6,12	6,85 7,33	7,72 8,20	8,41 8,85	8,94 9,34	9,35 9,73	9,68 10,02	9,93 10,25	10,12 10,42	10,27 10,55	10,36 10,64	10,46 10,71	10,51 10,75	10,54 10,77	10,55 10,78	10,55 10,77	10,54 10,76
350	4,22 2,16	7,83 8,77	10,22 11,58	11,68 12,68	12,56 13,25	13,08 13,59	13,39 13,79	13,57 13,91	13,65 13,95	13,68 13,95	13,67 13,91	13,62 13,84	13,55 13,75	13,47 13,65	13,37 13,54	13,26 13,41	13,14 13,29	13,02 13,16	12,89 13,03	12,77 12,90
250	15,76 18,84	20,12 23,31	20,86 21,60	20,56 20,60	20,03 19,96	19,49 19,46	19,00 19,02	18,55 18,60	18,13 18,20	17,75 17,82	17,38 17,45	17,04 17,10	16,70 16,76	16,39 16,43	16,08 16,12	15,79 15,83	15,51 15,55	15,25 15,28	14,99 15,02	14,74 14,77
150	58,84 58,27	41,14 33,96	33,00 29,24	28,85 27,10	26,44 25,62	24,83 24,42	23,62 23,39	22,64 22,49	21,80 21,67	21,05 20,94	20,38 20,27	19,76 19,66	19,20 19,11	18,68 18,59	18,20 18,11	17,75 17,67	17,33 17,26	16,94 16,87	16,57 16,51	16,23 16,17
50	19,61 20,80	26,78 31,83	28,85 31,85	28,85 30,38	28,04 28,75	26,96 27,24	25,85 25,89	24,77 24,69	23,78 23,63	22,87 22,69	22,03 21,84	21,27 21,08	20,58 20,40	19,94 19,77	19,36 19,20	18,82 18,67	18,32 18,18	17,86 17,73	17,43 17,31	17,03 16,92

Table III.3 KNMI-results.

Concentrations in $\mu g/m^3$

Z in m \ X in km	5	15	25	35	45	55	65	75	85	95	105	115	125	135	145	155	165	175	185	195
950	0,00 0,00	0,00 0,00	0,00 0,00	0,00 0,00	0,10 0,02	0,00 0,06	−0,20 0,12	0,10 0,22	0,60 0,34	0,70 0,48	0,60 0,64	0,70 0,80	0,90 0,98	1,20 1,15	1,30 1,33	1,50 1,50	1,70 1,68	1,90 1,85	2,10 2,01	2,30 2,17
850	0,00 0,00	0,00 0,00	0,00 0,00	0,10 0,02	0,10 0,09	0,10 0,21	0,00 0,39	0,40 0,59	1,10 0,82	1,40 1,07	1,30 1,32	1,40 1,57	1,80 1,82	2,00 2,05	2,30 2,28	2,50 2,50	2,70 2,71	2,90 2,91	3,10 3,10	3,30 3,28
750	0,00 0,00	0,00 0,00	0,10 0,03	0,20 0,14	0,40 0,37	0,50 0,69	0,60 1,04	1,20 1,42	2,10 1,79	2,50 2,16	2,40 2,50	2,60 2,83	3,10 3,14	3,40 3,42	3,70 3,69	3,90 3,93	4,10 4,16	4,40 4,37	4,60 4,56	4,70 4,73
650	0,00 0,00	0,10 0,02	0,20 0,21	0,70 0,65	1,20 1,24	1,60 1,85	1,90 2,45	2,70 3,00	3,90 3,50	4,40 3,95	4,30 4,36	4,50 4,72	5,00 5,05	5,30 5,34	5,60 5,59	5,80 5,82	6,00 6,03	6,20 6,21	6,40 6,37	6,50 6,52
550	0,00 0,00	0,30 0,24	1,10 1,13	2,20 2,28	3,20 3,33	3,90 4,23	4,30 4,98	5,30 5,61	6,60 6,14	7,10 6,59	6,90 6,96	7,00 7,28	7,50 7,56	7,80 7,78	8,00 7,98	8,10 8,14	8,30 8,28	8,40 8,39	8,50 8,49	8,60 8,57
450	0,00 0,12	1,80 1,86	4,00 4,34	5,90 6,12	7,20 7,33	7,80 8,20	8,10 8,85	9,00 9,34	10,30 9,73	10,60 10,02	10,20 10,25	10,10 10,42	10,50 10,55	10,70 10,64	10,70 10,71	10,80 10,75	10,80 10,77	10,80 10,78	10,80 10,77	10,80 10,75
350	3,00 2,16	7,40 8,77	10,90 11,58	12,50 12,68	13,30 13,25	13,30 13,59	13,10 13,79	13,60 13,91	14,60 13,95	14,70 13,95	13,90 13,91	13,60 13,84	13,70 13,75	13,70 13,65	13,60 13,54	13,50 13,41	13,30 13,29	13,20 13,16	13,10 13,03	12,90 12,90
250	14,00 18,64	21,60 23,31	22,10 21,60	20,90 20,60	20,20 19,96	19,30 19,46	18,40 19,02	18,30 18,60	19,00 18,20	18,70 17,82	17,50 17,45	16,80 17,10	16,80 16,76	16,50 16,43	16,20 16,12	15,90 15,83	15,60 15,55	15,30 15,28	15,10 15,02	14,80 14,77
150	32,00 58,27	37,80 33,96	30,90 29,24	27,60 27,10	26,10 25,62	24,40 24,42	22,80 23,39	22,30 22,49	22,60 21,67	21,90 20,94	20,40 20,27	19,40 19,66	19,20 19,11	18,70 18,59	18,30 18,11	17,80 17,67	17,40 17,26	17,00 16,87	16,60 16,51	16,30 16,17
50	17,00 20,80	28,60 31,83	31,70 31,85	30,60 30,38	29,20 28,75	27,20 27,24	25,40 25,89	24,50 24,69	24,70 23,63	23,70 22,69	22,00 21,84	20,90 21,08	20,50 20,40	19,90 19,77	19,40 19,20	18,80 18,67	18,30 18,18	17,80 17,73	17,40 17,31	17,00 16,92

Table III.4 RIV-results case A.

Concentrations in $\mu g/m^2$

Z in m \ X in km	5	15	25	35	45	55	65	75	85	95	105	115	125	135	145	155	165	175	185	195
950	0,00 0,00	0,00 0,00	0,00 0,00	0,00 0,00	0,00 0,03	0,00 0,06	0,00 0,12	0,00 0,22	0,00 0,34	0,00 0,48	0,00 0,64	0,00 0,80	0,00 0,98	0,00 1,15	0,00 1,33	0,00 1,50	0,00 1,68	0,00 1,85	0,00 2,01	0,00 2,17
850	0,00 0,00	0,00 0,00	0,00 0,00	0,00 0,02	0,00 0,09	0,50 0,21	1,00 0,39	1,00 0,59	1,00 0,82	1,50 1,07	2,00 1,32	2,00 1,57	2,50 1,82	2,50 2,05	2,50 2,28	3,00 2,50	3,00 2,71	3,50 2,91	3,50 3,10	3,50 3,28
750	0,00 0,00	0,00 0,00	0,00 0,03	0,50 0,14	1,00 0,37	1,00 0,69	1,50 1,04	2,00 1,43	2,50 1,79	2,50 2,16	3,50 2,50	3,50 2,83	3,50 3,14	4,00 3,42	4,50 3,69	4,50 3,92	4,50 4,16	4,50 4,37	5,00 4,56	5,50 4,73
650	0,00 0,00	0,00 0,02	1,00 0,21	1,50 0,65	2,50 1,24	2,50 1,85	3,50 2,45	3,50 3,00	4,50 3,50	4,50 3,95	5,50 4,36	5,50 4,72	5,50 5,05	6,00 5,34	6,50 5,59	6,50 5,82	6,50 6,03	6,50 6,21	6,50 6,37	7,00 6,52
550	0,50 0,00	1,50 0,24	3,00 1,13	4,00 2,28	5,00 3,33	5,50 4,23	6,00 4,98	7,00 5,61	7,00 6,14	7,50 6,59	8,00 6,96	8,00 7,28	8,50 7,56	8,50 7,78	8,50 7,98	8,50 8,14	9,50 8,28	8,50 8,39	8,50 8,49	9,50 8,57
450	3,00 0,12	5,50 1,86	7,50 4,34	8,50 6,12	9,00 7,33	9,50 8,20	9,50 8,85	10,00 9,34	11,00 9,73	11,00 10,02	11,00 10,25	11,00 10,42	11,00 10,55	11,00 10,64	11,00 10,71	11,50 10,75	11,50 10,77	11,50 10,78	11,50 10,77	11,50 10,75
350	11,00 2,16	13,00 8,77	14,00 11,58	14,50 12,68	14,50 13,25	14,50 13,59	14,50 13,79	15,00 13,91	14,00 13,95	14,00 13,95	14,50 13,91	14,00 13,84	14,00 13,75	14,00 13,65	13,50 13,54	13,50 13,41	13,50 13,29	13,50 13,16	13,50 13,03	13,50 12,90
250	25,50 18,64	22,50 23,31	22,00 21,60	21,00 20,60	20,50 19,96	19,50 19,46	19,00 19,02	19,00 18,60	18,00 18,20	17,50 17,82	17,50 17,45	17,00 17,10	16,50 16,76	16,50 16,43	15,50 16,12	15,50 15,83	15,50 15,55	15,50 15,28	15,00 15,02	14,50 14,77
150	33,00 58,27	29,00 33,96	28,00 29,24	26,00 27,10	25,00 25,62	23,50 24,42	23,50 23,39	22,00 22,49	20,50 21,67	20,50 20,94	19,50 20,27	19,50 19,66	18,50 19,11	18,50 18,59	17,50 18,11	17,50 17,67	17,00 17,26	16,50 16,87	16,50 16,51	16,00 16,17
50	30,50 20,80	31,00 31,83	30,00 31,85	28,50 30,38	27,50 28,75	25,50 27,24	24,50 25,89	23,50 24,69	22,00 23,63	21,50 22,69	21,00 21,84	20,00 21,03	20,00 20,40	19,00 19,77	19,00 19,20	18,00 18,67	18,00 18,19	17,00 17,73	17,00 17,31	17,00 16,92

Table III.5 TNO(SAI)-results case A.

Concentrations in $\mu g/m^3$

Z in m \ X in km	5	15	25	35	45	55	65	75	85	95	105	115	125	135	145	155	165	175	185	195
950	0,00 0,00	0,00 0,00	0,01 0,00	0,04 0,00	0,10 0,02	0,20 0,06	0,35 0,12	0,54 0,22	0,78 0,34	1,06 0,48	1,37 0,64	1,70 0,80	2,05 0,98	2,41 1,15	2,77 1,33	3,14 1,50	3,49 1,68	3,84 1,85	4,18 2,01	4,50 2,17
850	0,00 0,00	0,01 0,00	0,04 0,00	0,12 0,02	0,25 0,09	0,43 0,21	0,68 0,39	0,96 0,59	1,29 0,82	1,64 1,07	2,01 1,32	2,39 1,57	2,77 182	3,41 2,05	3,51 2,28	3,88 2,50	4,22 2,71	4,56 2,91	4,88 3,10	5,17 3,28
750	0,00 0,00	0,04 0,00	0,16 0,03	0,38 0,14	0,69 0,37	1,08 0,69	1,52 1,04	1,99 1,42	2,47 1,79	2,95 2,16	3,41 2,50	3,85 2,83	4,28 3,14	4,65 3,42	5,01 3,69	5,35 3,93	5,66 4,16	5,95 4,37	6,22 4,56	6,47 4,73
650	0,03 0,00	0,20 0,02	0,57 0,21	1,13 0,65	1,81 1,24	2,55 1,85	3,30 2,45	4,00 3,00	4,65 3,50	5,22 3,95	5,74 4,36	6,19 4,72	6,59 5,05	6,93 5,34	7,24 5,59	7,91 5,82	7,74 6,03	7,96 6,21	8,13 6,37	8,29 6,52
550	0,19 0,00	0,83 0,24	1,86 1,13	3,10 2,28	4,37 3,33	5,34 4,23	6,56 4,98	7,40 5,61	8,09 6,14	8,65 6,59	9,09 6,96	9,44 7,28	9,72 7,56	9,93 7,78	10,09 7,98	10,21 8,14	10,31 8,28	10,38 8,39	10,43 8,49	10,47 8,57
450	0,93 0,12	2,97 1,86	5,41 4,34	7,86 6,12	9,48 7,33	10,83 8,20	11,80 8,85	12,45 9,34	12,89 9,73	13,16 10,02	13,32 10,25	13,39 10,42	13,40 10,55	13,37 10,64	13,30 10,71	13,21 10,75	13,11 10,77	12,99 10,78	12,88 10,77	12,76 10,75
350	4,21 2,16	9,67 8,77	13,93 11,58	16,59 12,68	18,02 13,25	18,66 13,59	18,84 13,79	18,76 13,91	18,53 13,95	18,23 13,95	17,88 13,91	17,52 13,84	17,15 13,75	16,79 13,65	16,43 13,54	16,09 13,41	15,76 13,29	15,45 13,16	15,15 13,03	14,87 12,90
250	17,46 18,64	27,27 23,31	30,24 21,60	30,13 20,60	28,97 19,96	27,57 19,46	26,20 19,02	24,93 18,60	23,80 18,20	22,77 17,82	21,85 17,45	21,03 17,10	20,27 16,76	19,59 16,43	18,96 16,12	18,39 15,83	17,85 15,55	17,36 15,28	16,91 15,02	16,50 14,77
150	67,26 58,27	60,33 33,96	49,96 29,24	42,62 27,10	37,56 25,62	33,89 24,42	31,09 23,39	28,86 22,49	27,04 21,67	25,50 20,94	24,19 20,27	23,05 19,66	22,05 19,11	21,17 18,59	20,38 1811	19,66 17,67	19,01 17,26	18,42 16,87	17,87 16,51	17,38 16,17
50	44,89 20,80	66,28 31,83	53,50 31,85	44,86 30,38	39,11 28,75	35,03 27,24	31,97 25,89	29,57 24,69	27,62 23,63	25,99 22,69	24,61 21,84	23,42 21,08	22,37 20,40	21,45 19,77	20,63 19,20	19,89 18,67	19,22 18,18	18,61 17,73	18,04 17,31	17,55 16,92

Table III.6 Phase 1, Case A

		Analytical solution	
	correlation coefficient	lin. 0.99	log. 0.97
THD	rootmeansquare error	0.9039 μg/m^3	
	ratio of centre box integrated concentrations	0.99	
	number of grid points compared	183	
	correlation coefficient	lin. 0.97	log. 0.99
KNMI	rootmeansquare error	2.075 μg/m^3	
	ratio of centre box integrated concentrations	0.98	
	number of grid points compared	180	
	correlation coefficient	lin. 0.96	log. 0.98
RIV	rootmeansquare error	2.480 μg/m^3	
	ratio of centre box integrated concentrations	1.02	
	number of grid points compared	169	
	correlation coefficient	lin. 0.96	log. 0.98
TNO (SAI)	rootmeansquare error	5.587 μg/m^3	
	ratio of centre box integrated concentrations	1.30	
	number of grid points compared	183	

Table III.7 Vertically integrated concentrations in the plume centre Phase I, Case A ($\mu g/m^3$)

model	Columns 1	2	3	4	5	6	7	8	9	10	11	12	13	14	15	16	17	18	19	20	total
THD	99	99	99	99	99	99	99	99	99	98	98	98	97	97	96	96	95	95	94	94	1962
KNMI	66	97	101	100	101	98	94	97	105	105	99	96	99	99	99	98	98	97	97	97	1951
TNO (SAI)	134	167	155	146	140	135	132	129	127	125	123	121	120	119	118	117	116	115	114	113	2577
RIV	103	102	105	104	105	102	102	103	100	100	102	100	100	100	97	98	98	97	97	98	2018
ANALYT. SOL.	99	99	99	99	99	99	99	99	99	99	99	99	99	99	98	98	98	97	97	96	1982

tration in the plume centre indicated that there is a wavelike pattern dependent on distance in the values of the columns. The RIV-model seems to possess too high concentrations in the plume centre. The same is true for the TNO (SAI)-model.

There is a high correlation between the results of the models and the ones of the analytical model. The root means square error is smallest for the THD-model and largest for the TNO(SAI)-model.

Case B

For case B a similar form of presentation of the results was chosen. Tables III.8, III.9, III.10 and III.11 show the results of the four models, in mg/m^3, plotted above the values of the analytical solution. Additional information is shown in tables III.12 and III.13. The best agreement occurs again between the results of the THD-model and the analytical solution. The results presented in tables III.12 and III.13 have been obtained by using only the concentrations calculated upto x = 95 km because in the KNMI-calculation a restricted area has been used. The agreement between calculated concentrations and the analytical solution improves with distance.

In this comparison the results of the TNO (SAI)-model show smaller concentrations in the plume centre than those of the analytical model.

The results of the four models do not show the same tendency of the vertically integrated concentration for a decrease with distance from the source as in case A.

Case C

Figures III.1, III.2, III.3 and III.4 show the concentration fields of the four models varying with time. The tables III.14, III.15 and III.16 give an impression of the comparability of the results. There are three facts which immediately call for attention:

1. The KNMI-model does not have any results for the first hour. This is due to the fact that the model produces only values below the height of the inversion. At the first hour the inversion is at 200 m and the source is at 300 m height.
2. The differences in the treatment of vertical diffusion in the various models introduce large RMS-errors for the first hour (see table III.15).
3. The poor agreement of the results of the TNO (SAI)-model for the second field is due to a coding error in the source height occurring only at the moment that the mixing height equals the source height. For the calculations an old, not corrected, version of the model has been used.

Table III.8 THD-results case B.

Concentrations in mg/m^3

Z in m \ X in km	5	15	25	35	45	55	65	75	85	95	105	115	125	135	145	155	165	175	185	195
95	0,27 / 0,08	0,56 / 0,54	0,76 / 0,83	0,87 / 0,93	0,93 / 0,98	0,96 / 0,99	0,97 / 1,00	0,98 / 1,00	0,98 / 1,00	0,98 / 1,00	0,98 / 1,00	0,98 / 1,00	0,97 / 1,00	0,97 / 1,00	0,97 / 1,00	0,97 / 1,00	0,97 / 0,99	0,96 / 0,99	0,96 / 0,99	0,96 / 0,99
85	0,30 / 0,11	0,59 / 0,58	0,78 / 0,84	0,88 / 0,94	0,93 / 0,98	0,96 / 0,99	0,97 / 1,00	0,98 / 1,00	0,98 / 1,00	0,98 / 1,00	0,98 / 1,00	0,98 / 1,00	0,97 / 1,00	0,97 / 1,00	0,97 / 1,00	0,97 / 1,00	0,97 / 0,99	0,96 / 0,99	0,96 / 0,99	0,96 / 0,99
75	0,36 / 0,18	0,65 / 0,67	0,82 / 0,88	0,90 / 0,95	0,94 / 0,98	0,96 / 0,99	0,97 / 1,00	0,98 / 1,00	0,98 / 1,00	0,98 / 1,00	0,98 / 1,00	0,98 / 1,00	0,97 / 1,00	0,97 / 1,00	0,97 / 1,00	0,97 / 1,00	0,97 / 0,99	0,96 / 0,99	0,96 / 0,99	0,96 / 0,99
65	0,45 / 0,30	0,74 / 0,78	0,87 / 0,92	0,93 / 0,97	0,96 / 0,99	0,97 / 1,00	0,98 / 1,00	0,98 / 1,00	0,98 / 1,00	0,98 / 1,00	0,98 / 1,00	0,98 / 1,00	0,97 / 1,00	0,97 / 1,00	0,97 / 1,00	0,97 / 1,00	0,97 / 0,99	0,96 / 0,99	0,96 / 0,99	0,96 / 0,99
55	0,59 / 0,48	0,85 / 0,92	0,94 / 0,97	0,97 / 0,99	0,98 / 1,00	0,98 / 1,00	0,98 / 1,00	0,98 / 1,00	0,98 / 1,00	0,98 / 1,00	0,98 / 1,00	0,98 / 1,00	0,97 / 1,00	0,97 / 1,00	0,97 / 1,00	0,97 / 1,00	0,97 / 0,99	0,96 / 0,99	0,96 / 0,99	0,96 / 0,99
45	0,79 / 0,74	0,99 / 1,06	1,02 / 1,03	1,01 / 1,01	1,00 / 1,00	0,99 / 1,00	0,99 / 1,00	0,98 / 1,00	0,98 / 1,00	0,98 / 1,00	0,98 / 1,00	0,98 / 1,00	0,97 / 1,00	0,97 / 1,00	0,97 / 1,00	0,97 / 1,00	0,97 / 0,99	0,96 / 0,99	0,96 / 0,99	0,96 / 0,99
35	1,07 / 1,11	1,16 / 1,21	1,10 / 1,08	1,05 / 1,03	1,02 / 1,01	1,00 / 1,00	0,99 / 1,00	0,99 / 1,00	0,98 / 1,00	0,98 / 1,00	0,98 / 1,00	0,98 / 1,00	0,97 / 1,00	0,97 / 1,00	0,97 / 1,00	0,97 / 1,00	0,97 / 0,99	0,96 / 0,99	0,96 / 0,99	0,96 / 0,99
25	1,45 / 1,62	1,33 / 1,33	1,17 / 1,12	1,08 / 1,05	1,04 / 1,02	1,01 / 1,01	1,00 / 1,00	0,99 / 1,00	0,98 / 1,00	0,98 / 1,00	0,98 / 1,00	0,98 / 1,00	0,97 / 1,00	0,97 / 1,00	0,97 / 1,00	0,97 / 1,00	0,97 / 0,99	0,96 / 0,99	0,96 / 0,99	0,96 / 0,99
15	1,98 / 2,32	1,49 / 1,43	1,23 / 1,16	1,11 / 1,06	1,05 / 1,02	1,02 / 1,01	1,00 / 1,00	0,99 / 1,00	0,99 / 1,00	0,98 / 1,00	0,98 / 1,00	0,98 / 1,00	0,97 / 1,00	0,97 / 1,00	0,97 / 1,00	0,97 / 1,00	0,97 / 0,99	0,96 / 0,99	0,96 / 0,99	0,96 / 0,99
5	2,71 / 3,06	1,60 / 1,47	1,26 / 1,17	1,12 / 1,06	1,05 / 1,02	1,02 / 1,01	1,00 / 1,00	0,99 / 1,00	0,99 / 1,00	0,98 / 1,00	0,98 / 1,00	0,98 / 1,00	0,97 / 1,00	0,97 / 1,00	0,97 / 1,00	0,97 / 1,00	0,97 / 0,99	0,96 / 0,99	0,96 / 0,99	0,96 / 0,99

Table III.9 KNMI-results case B.

X in km

Z in m	5	15	25	35	45	55	65	75	85	95
95	0,08 0,08	0,39 0,54	0,70 0,83	0,90 0,93	0,96 0,98	1,00 0,99	1,00 1,00	1,00 1,00	0,99 1,00	1,00 1,00
85	0,10 0,11	0,42 0,58	0,73 0,84	0,91 0,94	0,96 0,98	1,00 0,99	1,00 1,00	1,00 1,00	0,99 1,00	1,00 1,00
75	0,14 0,18	0,49 0,67	0,78 0,88	0,93 0,95	0,97 0,98	1,00 0,99	0,98 1,00	1,00 1,00	1,00 1,00	1,00 1,00
65	0,21 0,30	0,60 0,78	0,85 0,92	0,96 0,97	0,98 0,99	1,00 1,00	1,00 1,00	1,00 1,00	1,00 1,00	1,00 1,00
55	0,31 0,48	0,75 0,92	0,94 0,97	0,99 0,99	0,99 1,00	1,01 1,00	1,00 1,00	1,00 1,00	1,00 1,00	1,00 1,00
45	0,45 0,74	0,92 1,06	1,04 1,03	1,03 1,01	1,01 1,00	1,01 1,00	1,00 1,00	1,00 1,00	1,00 1,00	1,00 1,00
35	0,64 1,11	1,13 1,21	1,14 1,08	1,06 1,03	1,02 1,01	1,02 1,00	1,01 1,00	1,00 1,00	1,00 1,00	1,00 1,00
25	0,89 1,62	1,37 1,33	1,22 1,12	1,08 1,05	1,03 1,02	1,02 1,01	1,01 1,00	1,00 1,00	1,00 1,00	1,00 1,00
15	1,11 2,32	1,55 1,43	1,29 1,16	1,10 1,06	1,04 1,02	1,02 1,01	1,01 1,00	1,01 1,00	1,00 1,00	1,00 1,00
5	1,14 3,06	1,61 1,47	1,32 1,17	1,11 1,06	1,04 1,02	1,02 1,01	1,01 1,00	1,01 1,00	1,00 1,00	1,00 1,00

Concentrations in mg/m^3

Summary and conclusions of phase I

The diversity of the techniques used in the different models reflects the fact that they were developed independently and for different purposes. This diversity leads (not at all surprisingly) to differences in the model outputs. The results of phase I show that all the models are able to simulate well defined and relatively simple situations (see Table III.1). This is in spite of the fact that the structure of the model is different.

In test case A (see Table III.6) there is an excellent agreement between the numerical and the analytical solution. The vertically integrated concentration shows, in the case of the TNO (SAI)-model, considerably larger values for short distances

(text continued on page 200)

Table III.10 RIV-results case B.

Concentrations in mg/m³

Z in m \ X in km	5	15	25	35	45	55	65	75	85	95	105	115	125	135	145	155	165	175	185	195
95	0,09 0,08	0,55 0,54	0,83 0,83	0,93 0,93	0,98 0,98	0,99 0,99	0,99 1,00	1,01 1,00	1,00 1,00	0,99 1,00	1,00 1,00	0,99 1,00	1,00 1,00	0,99 1,00	1,00 1,00	0,99 1,00	1,02 0,99	0,99 0,99	0,98 0,99	1,00 0,99
85	0,13 0,11	0,60 0,58	0,85 0,84	0,93 0,94	0,98 0,98	0,99 0,99	0,99 1,00	1,01 1,00	1,00 1,00	0,99 1,00	1,00 1,00	0,99 1,00	1,00 1,00	0,99 1,00	1,00 1,00	0,99 1,00	1,02 0,99	0,99 0,99	0,98 0,99	1,00 0,99
75	0,20 0,18	0,68 0,67	0,88 0,88	0,95 0,95	0,99 0,98	0,99 0,99	0,99 1,00	1,01 1,00	1,00 1,00	0,99 1,00	1,00 1,00	0,99 1,00	1,00 1,00	0,99 1,00	1,00 1,00	0,99 1,00	1,02 0,99	0,99 0,99	0,98 0,99	1,00 0,99
65	0,31 0,30	0,80 0,78	0,93 0,92	0,96 0,97	0,99 0,99	0,99 1,00	0,99 1,00	1,01 1,00	1,00 1,00	0,99 1,00	1,00 1,00	0,99 1,00	1,00 1,00	0,99 1,00	1,00 1,00	0,99 1,00	1,02 0,99	0,99 0,99	0,98 0,99	1,00 0,99
55	0,48 0,48	0,93 0,92	0,98 0,97	0,98 0,99	1,00 1,00	0,99 1,00	0,99 1,00	1,01 1,00	1,00 1,00	0,99 1,00	1,00 1,00	0,99 1,00	1,00 1,00	0,99 1,00	1,00 1,00	0,99 1,00	1,02 0,99	0,99 0,99	0,98 0,99	1,00 0,99
45	0,68 0,74	1,07 1,06	1,03 1,03	1,00 1,01	1,01 1,00	0,99 1,00	0,99 1,00	1,01 1,00	1,00 1,00	0,99 1,00	1,00 1,00	0,99 1,00	1,00 1,00	0,99 1,00	1,00 1,00	0,99 1,00	1,02 0,99	0,99 0,99	0,98 0,99	1,00 0,99
35	0,92 1,11	1,21 1,21	1,08 1,08	1,02 1,03	1,01 1,01	0,99 1,00	0,99 1,00	1,01 1,00	1,00 1,00	0,99 1,00	1,00 1,00	0,99 1,00	1,00 1,00	0,99 1,00	1,00 1,00	0,99 1,00	1,02 0,99	0,99 0,99	0,98 0,99	1,00 0,99
25	1,14 1,62	1,33 1,33	1,13 1,12	1,04 1,05	1,02 1,02	0,99 1,01	0,99 1,00	1,01 1,00	1,00 1,00	0,99 1,00	1,00 1,00	0,99 1,00	1,00 1,00	0,99 1,00	1,00 1,00	0,99 1,00	1,02 0,99	0,99 0,99	0,98 0,99	1,00 0,99
15	1,32 2,32	1,42 1,43	1,15 1,16	1,05 1,06	1,03 1,02	0,99 1,01	0,99 1,00	1,01 1,00	1,00 1,00	0,99 1,00	1,00 1,00	0,99 1,00	1,00 1,00	0,99 1,00	1,00 1,00	0,99 1,00	1,02 0,99	0,99 0,99	0,98 0,99	1,00 0,99
5	1,42 3,06	1,46 1,47	1,17 1,17	1,06 1,06	1,03 1,02	0,99 1,01	0,99 1,00	1,01 1,00	1,00 1,00	0,99 1,00	1,00 1,00	0,99 1,00	1,00 1,00	0,99 1,00	1,00 1,00	0,99 1,00	1,02 0,99	0,99 0,99	0,98 0,99	1,00 0,99

Table III.11 TNO(SAI)-results case B.

Concentrations in mg/m^3

Z in m \ X in km	5	15	25	35	45	55	65	75	85	95	105	115	125	135	145	155	165	175	185	195
95	0,22 0,08	0,50 0,54	0,71 0,83	0,33 0,93	0,88 0,98	0,88 0,99	0,87 1,00	0,86 1,00	0,84 1,00	0,83 1,00	0,82 1,00	0,81 1,00	0,80 1,00	0,79 1,00	0,78 1,00	0,78 1,00	0,77 0,99	0,76 0,99	0,75 0,99	0,75 0,99
85	0,25 0,11	0,53 0,58	0,73 0,84	0,84 0,94	0,88 0,98	0,88 0,99	0,87 1,00	0,86 1,00	0,84 1,00	0,83 1,00	0,82 1,00	0,81 1,00	0,80 1,00	0,79 1,00	0,78 1,00	0,78 1,00	0,77 0,99	0,76 0,99	0,75 0,99	0,75 0,99
75	0,29 0,18	0,59 0,67	0,78 0,88	0,86 0,95	0,89 0,98	0,88 0,99	0,87 1,00	0,86 1,00	0,84 1,00	0,83 1,00	0,82 1,00	0,81 1,00	0,80 1,00	0,79 1,00	0,78 1,00	0,78 1,00	0,77 0,99	0,76 0,99	0,75 0,99	0,75 0,99
65	0,37 0,30	0,68 0,78	0,93 0,92	0,89 0,97	0,89 0,99	0,88 1,00	0,87 1,00	0,86 1,00	0,84 1,00	0,83 1,00	0,82 1,00	0,81 1,00	0,80 1,00	0,79 1,00	0,78 1,00	0,78 1,00	0,77 0,99	0,76 0,99	0,75 0,99	0,75 0,99
55	0,48 0,48	0,87 0,92	0,91 0,97	0,92 0,99	0,90 1,00	0,89 1,00	0,87 1,00	0,86 1,00	0,84 1,00	0,83 1,00	0,82 1,00	0,81 1,00	0,80 1,00	0,79 1,00	0,78 1,00	0,78 1,00	0,77 0,99	0,76 0,99	0,75 0,99	0,75 0,99
45	0,65 0,74	0,95 1,06	0,99 1,03	0,95 1,01	0,91 1,00	0,89 1,00	0,87 1,00	0,86 1,00	0,84 1,00	0,83 1,00	0,82 1,00	0,81 1,00	0,80 1,00	0,79 1,00	0,78 1,00	0,78 1,00	0,77 0,99	0,76 0,99	0,75 0,99	0,75 0,99
35	0,90 1,11	1,13 1,21	1,09 1,08	0,97 1,03	0,92 1,01	0,89 1,00	0,87 1,00	0,85 1,00	0,84 1,00	0,83 1,00	0,82 1,00	0,81 1,00	0,80 1,00	0,79 1,00	0,78 1,00	0,78 1,00	0,77 0,99	0,76 0,99	0,75 0,99	0,75 0,99
25	1,28 1,62	1,31 1,33	1,11 1,12	0,98 1,05	0,92 1,02	0,89 1,01	0,87 1,00	0,85 1,00	0,84 1,00	0,83 1,00	0,82 1,00	0,81 1,00	0,80 1,00	0,79 1,00	0,78 1,00	0,78 1,00	0,77 0,99	0,76 0,99	0,75 0,99	0,75 0,99
15	1,81 2,32	1,45 1,43	1,14 1,16	0,99 1,06	0,92 1,02	0,89 1,01	0,87 1,00	0,85 1,00	0,84 1,00	0,83 1,00	0,82 1,00	0,81 1,00	0,80 1,00	0,79 1,00	0,78 1,00	0,78 1,00	0,77 0,99	0,76 0,99	0,75 0,99	0,75 0,99
5	1,73 3,06	1,50 1,47	1,15 1,17	0,99 1,06	0,92 1,02	0,89 1,01	0,87 1,00	0,85 1,00	0,84 1,00	0,83 1,00	0,82 1,00	0,81 1,00	0,80 1,00	0,79 1,00	0,78 1,00	0,78 1,00	0,77 0,99	0,76 0,99	0,75 0,99	0,75 0,99

Table III.12 Phase 1, Case B.

		Analytical solution	
	correlation coefficient	lin. 0.98	log. 0.95
THD	rootmeansquare error	0.071 mg/m^3	
	ratio of centre box integrated concentrations	0.99	
	number of grid points compared	100	
	correlation coefficient	lin. 0.66	log. 0.92
KNMI	rootmeansquare error	0.252 mg/m^3	
	ratio of centre box integrated concentrations	1.02	
	number of grid points compared	100	
	correlation coefficient	lin. 0.82	log. 0.97
RIV	rootmeansquare error	0.198 mg/m^3	
	ratio of centre box integrated concentrations	1.03	
	number of grid points compared	100	
	correlation coefficient	lin. 0.90	log. 0.91
TNO (SAI)	rootmeansquare error	0.196 mg/m^3	
	ratio of centre box integrated concentrations	0.86	
	number of grid points compared	100	

Table III.13 Vertically integrated concentrations in the plume centre. Phase I, Case B (mg/m³).

model	Columns 1	2	3	4	5	6	7	8	9	10	total
THD	9	9	9	9	9	9	9	9	9	9	90
KNMI	5	9	10	10	10	10	10	10	9	10	93
TNO (SAI)	8	9	9	8	9	8	8	8	8	8	83
RIV	6	10	10	9	10	9	9	10	10	9	92
ANALYT. SOL.	10	9	9	9	9	9	9	9	9	9	91

which decrease afterwards. This mass increase is caused by the use of the non-mass conservative form of the SHASTA algorithm which was incorporated in the Airshed Model based on tests for an urban area with distributed sources. In cases with large gradients, such as those examined in this study where the plume had a characteristic lateral dimension smaller than one grid cell, the conservative formulation would have been the more appropriate choice. In addition, the code also contains tests to insure that concentrations do not become negative. Again, this was a provision added to the urban photochemical model to treat situations where large emissions of NO might react very rapidly with O_3 yielding small negative concentrations. This leads to numerical problems when the chemical reaction rates are integrated. The mass conservation results might have been somewhat better if the background level had been set to a number other than zero.

The test case B shows good agreement with the exception of the KNMI-model. This can be explained by the smearing out of the source over a few grid points which results in this case in much lower concentrations close to the source.

(text continued on page 208)

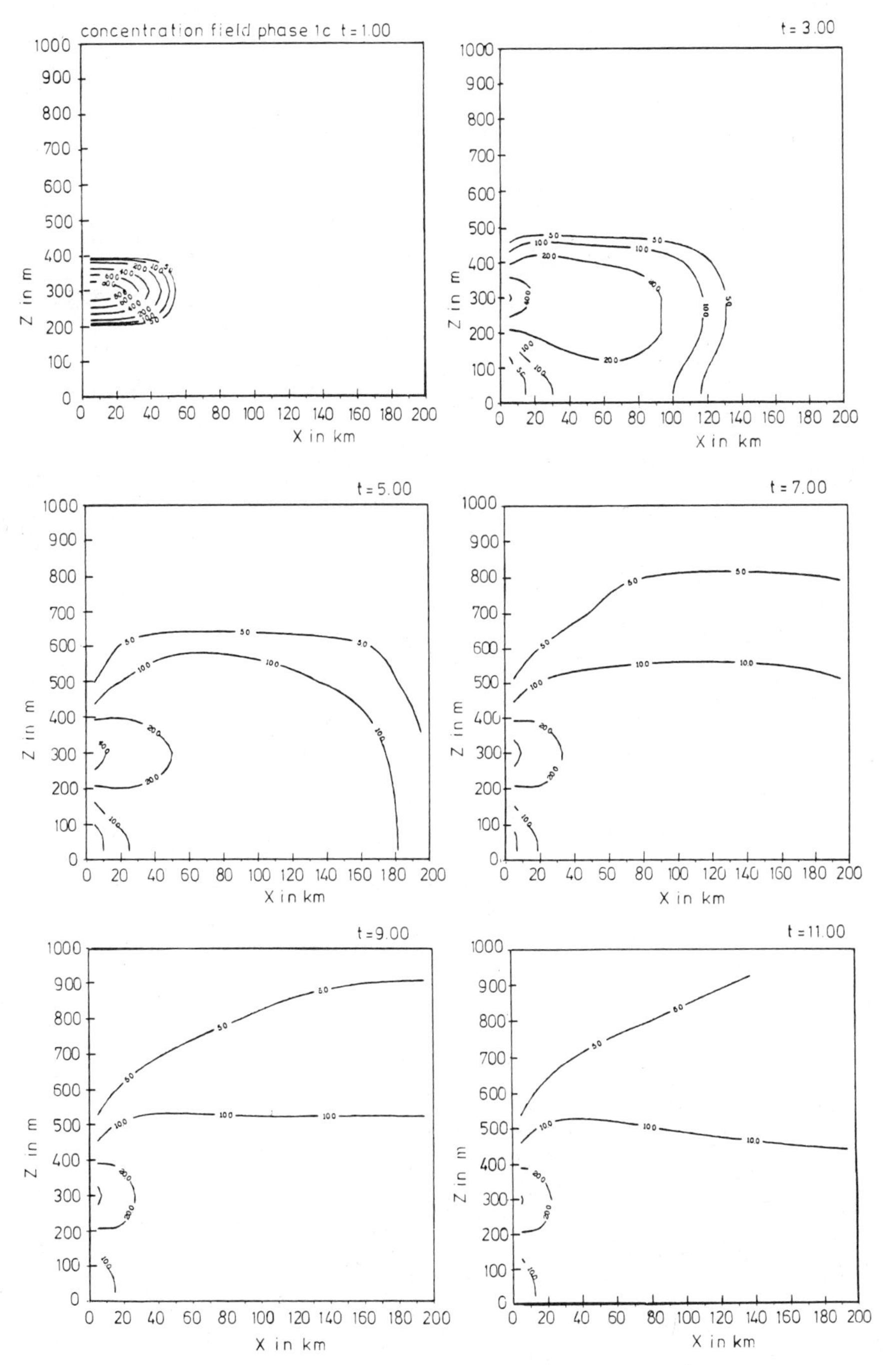

Fig. III.1 THD-results case C μg/m³.

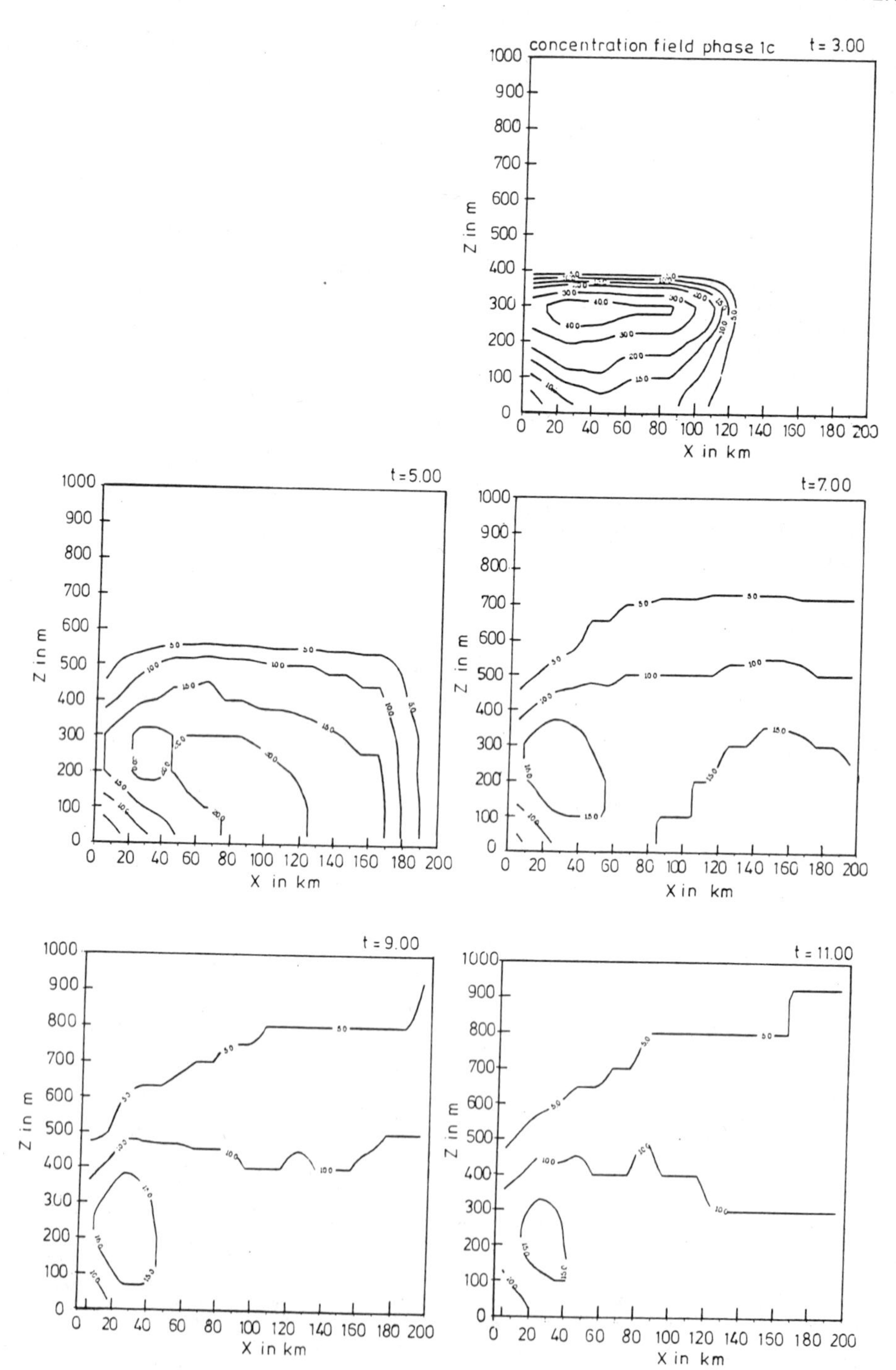

Fig. III.2 KNMI-results case C μg/m³.

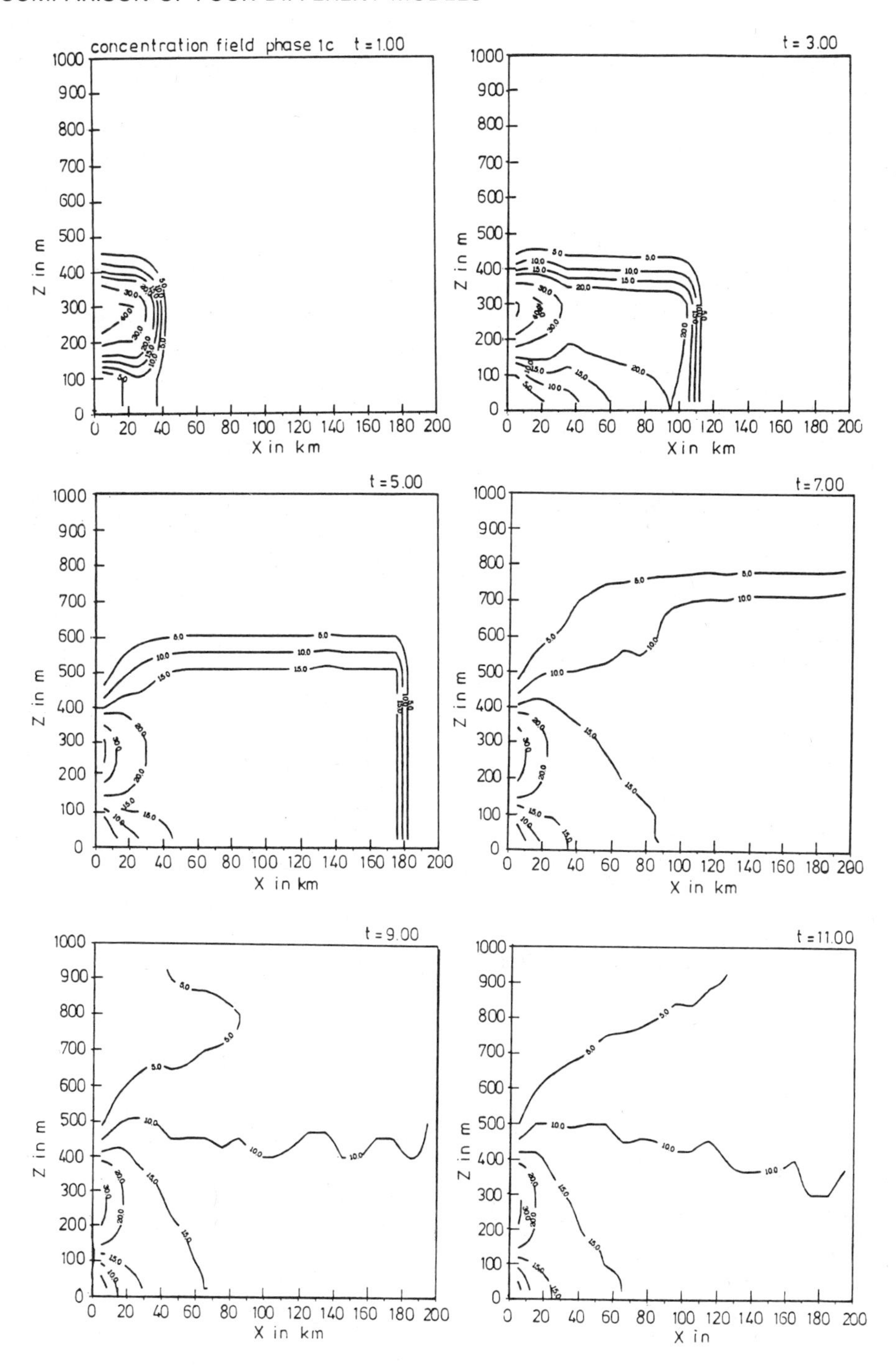

Fig. III.3 RIV-results case C μg/m³.

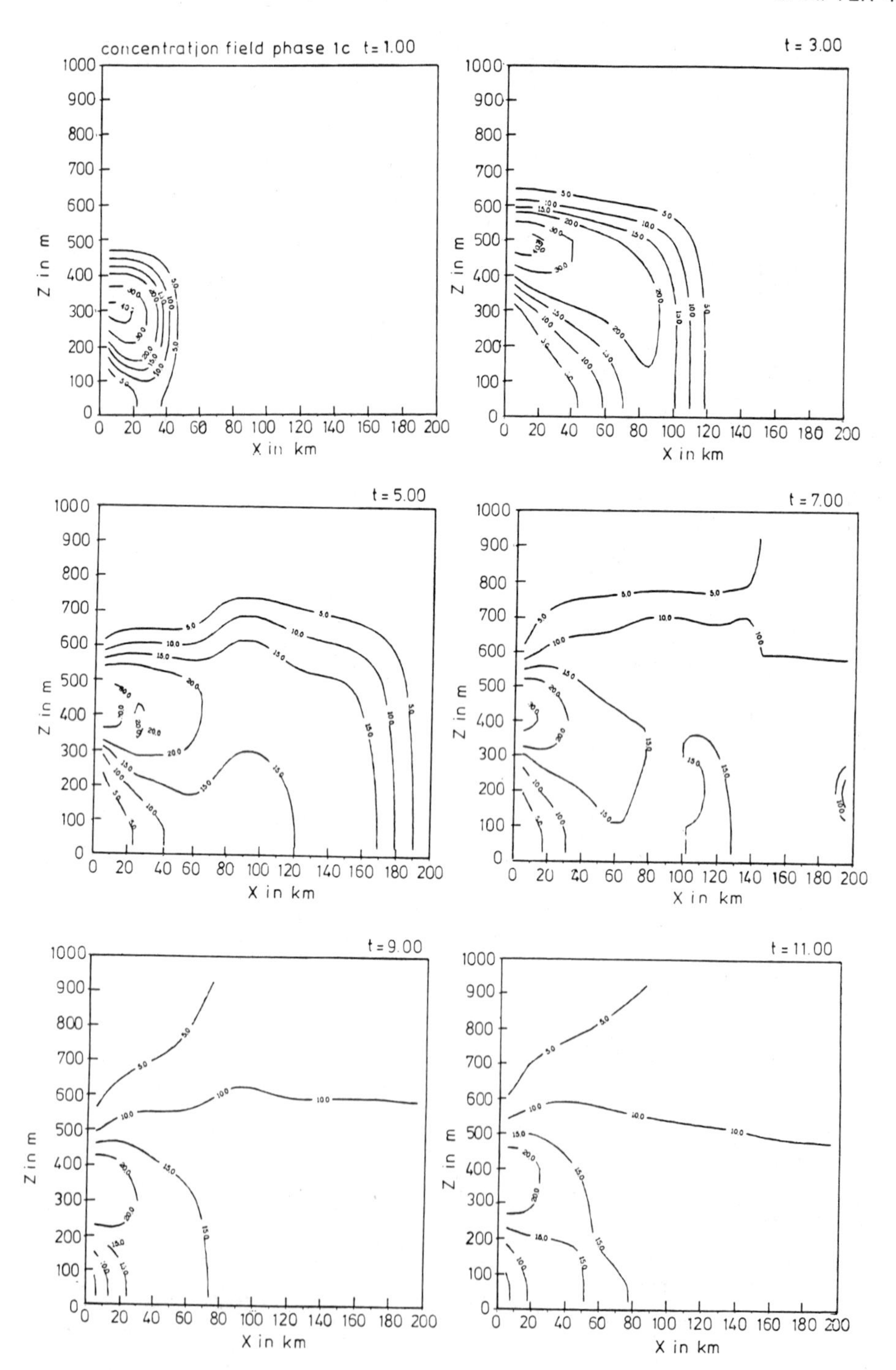

Fig. III.4 TNO(SAI)-results case C μg/m³.

Table III.14 Phase I, Case C, Correlation Coefficients.

		THD	KNMI		RIV		TNO (SAI)	
			lin.	log.	lin.	log.	lin.	log.
THD	1.00		-	-	0.74	0.74	0.83	0.83
	3.00		0.81	0.87	0.77	0.71	0.45	0.72
	5.00		0.72	0.90	0.80	0.80	0.44	0.50
	7.00		0.76	0.92	0.89	0.90	0.35	0.49
	9.00		0.86	0.96	0.90	0.85	0.95	0.98
	11.00		0.84	0.94	0.95	0.97	0.90	0.97
			0.80	0.92	0.84	0.83	0.65	0.75
KNMI	1.00				-	-	-	-
	3.00				0.72	0.74	0.49	0.55
	5.00				0.40	0.68	0.26	0.40
	7.00				0.66	0.83	0.24	0.34
	9.00				0.81	0.85	0.88	0.94
	11.00				0.85	0.93	0.81	0.88
					0.69	0.81	0.54	0.62
RIV	1.00						0.88	0.87
	3.00						-0.11	0.20
	5.00						0.23	0.36
	7.00						0.12	0.23
	9.00						0.87	0.86
	11.00						0.85	0.92
							0.47	0.57
TNO (SAI)	1.00							
	3.00							
	5.00							
	7.00							
	9.00							
	11.00							

Table III.15 Phase I, Case C, Rootmeansquare errors (μg/m³).

		THD	KNMI	RIV	TNO (SAI)
	1.00		-	(45.9)*	(41.0)*
	3.00		7.8	6.3	11.1
	5.00		4.5	3.8	5.9
THD	7.00		3.5	2.5	6.7
	9.00		2.9	2.3	1.7
	11.00		2.8	1.5	2.1
			4.3	3.3	5.5
	1.00			-	-
	3.00			8.3	14.4
	5.00			5.1	6.2
KNMI	7.00			3.7	7.0
	9.00			2.8	2.9
	11.00			2.4	3.0
				4.5	6.7
	1.00				(5.3)*
	3.00				13.8
	5.00				6.1
RIV	7.00				7.3
	9.00				2.5
	11.00				2.5
					6.4
	1.00				
	3.00				
	5.00				
TNO	7.00				
(SAI)	9.00				
	11.00				

* For the calculation of the mean RMS-error the RMS-values in the brackets were not considered

Table III.16 Vertically integrated concentrations at the plume centre, phase I, case C ($\mu g/m^3$).

model		Columns 1	2	3	4	5	6	7	8	9	10	11	12	13	14	15	16	17	18	19	20	total	exact value
THD	1.00	99	95	82	54	23	4															359.9	360
	3.00	100	102	104	105	106	106	105	100	91	77	59	40	24	12	5	1					1145	1080
	5.00	101	103	104	106	107	107	107	108	108	108	108	107	105	101	95	86	75	61	47	34	1889	1800
	7.00	102	104	105	106	107	107	107	107	107	107	107	107	107	107	107	107	106	105	103	100	2128	
	9.00	102	104	105	106	106	105	105	104	104	103	103	103	103	102	102	102	102	102	102	102	2079	
	11.00	102	105	105	105	104	103	102	101	100	99	98	98	97	96	96	95	95	94	94	94	1996	
KNMI	1.00	-	-	-	-	-	-	-	-	-	-	-	-	-	-	-	-	-	-	-	-	-	360
	3.00	71	87	102	104	106	97	91	91	91	74	56	31	6	2							1010	1080
	5.00	50	68	86	94	99	105	109	108	113	110	107	103	99	93	90	84	83	57	33	15	1706	1800
	7.00	51	70	87	90	92	89	93	91	92	93	06	99	104	107	113	113	108	106	106	101	1902	
	9.00	54	76	100	97	91	93	90	93	94	92	93	92	94	91	91	92	94	96	97	103	1823	
	11.00	53	69	86	89	83	86	85	82	101	96	94	90	83	84	87	87	88	89	87	87	1722	
RIV	1.00	98	93	99	56																	346.3	360
	3.00	105	105	96	79	86	90	93	93	95	97	76										1019	1080
	5.00	101	99	98	101	104	104	103	104	104	101	101	101	101	107	101	101	101	101			1841	1800
	7.00	99	98	99	100	102	103	103	102	102	102	103	102	102	104	103	103	103	102	102	103	2045	
	9.00	99	99	99	101	99	100	98	98	100	97	98	98	100	100	98	97	100	100	98	100	1989	
	11.00	99	100	100	100	99	101	97	95	95	92	92	93	90	90	90	88	89	88	88	89	1885	
TNO (SAI)	1.00	80	100	92	58	21	2															356.8	360
	3.00	82	95	99	102	105	109	116	122	120	102	70	35	11	1							1177	1080
	5.00	90	104	109	111	113	114	113	111	112	112	111	114	112	110	107	107	108	108	107	95	2179	1800
	7.00	90	103	97	110	113	113	112	110	109	109	110	111	115	119	122	116	99	71	41	17	1997	
	9.00	93	111	116	118	118	117	114	112	110	109	109	109	108	108	107	107	107	108	108	107	2204	
	11.00	93	108	112	113	112	111	110	109	108	107	106	105	104	102	100	99	98	98	98	98	2100	

The test case C is different from the other two test cases insofar that it simulates a nonstationary meteorological situation for which no analytical solution exists. Therefore only model results against other model results can be compared. Here the TNO (SAI)-model shows poor correlations with the others at 03.00 hrs, because of the deficiency in the source height already discussed in paragraph d). The agreement between the different models is less good for the first-stage of the fumigation process. This reflects also the different handling of the sources in the models.

Obviously, the RIV- and TNO (SAI)-model show already a downward mixing at the first hour. In these models the mixing height jumpes at the half hour. Consequently, already during the second hour the mixing height reaches the box where the pollutants are emitted. This explains the large RMS-value for these models in comparison with THD-model at 01.00 hrs (see Table III.15).

Notwithstanding the foregoing mentioned discrepancies, the numerical accuracy of all models is judged more than adequate for all practical applications.

IV. DATA BASES

IV.a Emission data base

Introduction

In the definition phase of the project it was decided that runs should be carried out using a real emission inventory for SO_2 of the area considered. However, because it was also decided that no attention would be given to boundary conditions (in fact, for the calculations the boundary conditions were set to zero) it seemed appropriate to use only a yearly averaged inventory supplemented with some monthly information concerning space-heating. So, - although in principle possible - an episodic specific inventory has not been set up.

Source description

Two Source Categories are discerned:

a) Point sources

These comprise power plants and large industrial emitters. They are assumed to emit continuously.

b) Surface sources

Surface sources have the dimensions of a grid cell, i.e. 10 x 10 km^2.

They are divided into two sub-categories:

- Industrial surface sources, supposed to emit every week from Monday to Friday and from 8.00 to 16.00 hours, i.e. 2000 hours/a.
- Space-heating sources, whose emissions have a seasonal variation that is assumed to follow the pattern as given in table IV.1.

Table IV.1 Seasonal variation of space-heating emissions

	% of full capacity	% of total yearly emission
Sept.	25	5
Oct.	50	10
Nov.	50	10
Dec.	75	15
Jan.	100	20
Febr.	100	18
March	60	12
April	50	10

They emit every day of the months indicated in table IV.1. Whether it is assumed that any space-heating installation emits every hour in these months at levels as given or that any installation, when operative, emits at full capacity but only during a fraction of the time, in both cases the monthly emission will be the same-percentage of the yearly total. Surface sources have an assumed geometric height of 20 m. Emissions are estimated for 1980. Figures IV.1 and IV.2 show the emission data divided into point source and surface source emissions. The sum of both is shown in Figure IV.3.

Federal Republic of Germany

Emissions from the Mid-Western part of the FRG were taken from C. Veldt (1982). Here the Ruhr-area and the Köln-Bonn region are dominant and emissions from these so-called "Belastungsgebiete" were directly taken from inventories published between 1976 and 1982 ("Luftreinhaltepläne", MAGS, Düsseldorf). Point source emissions outside the "Belastungsgebiete" - mainly powerplants - were estimated with data from Glückauf GmbH (1982) and emission factors from UBA (1980).

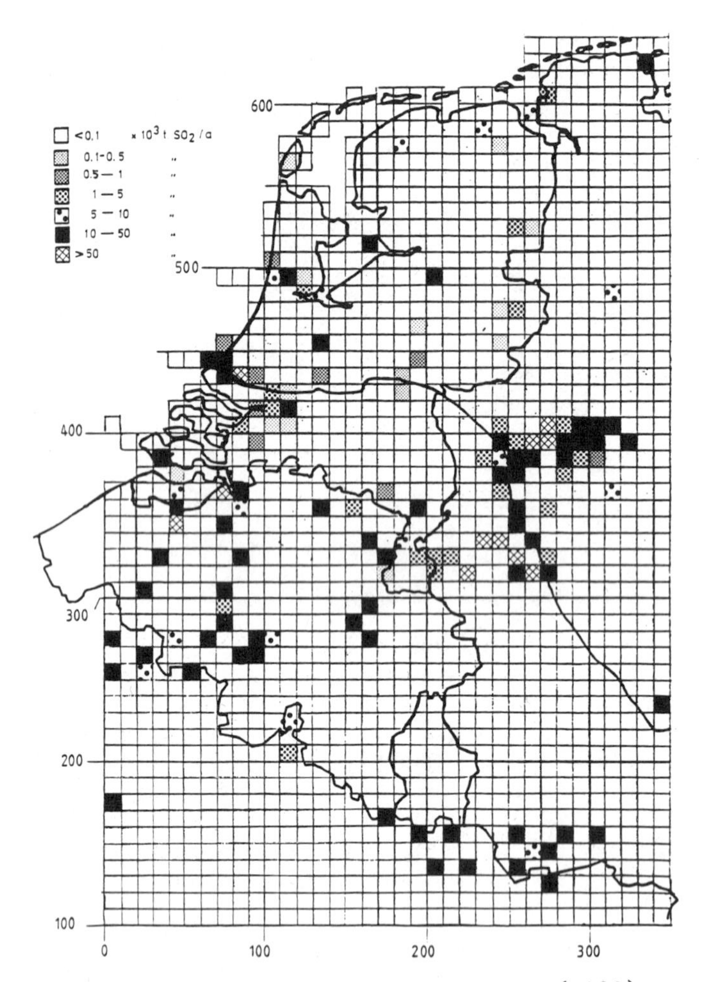

Fig. IV.1 Point source emissions (1980).

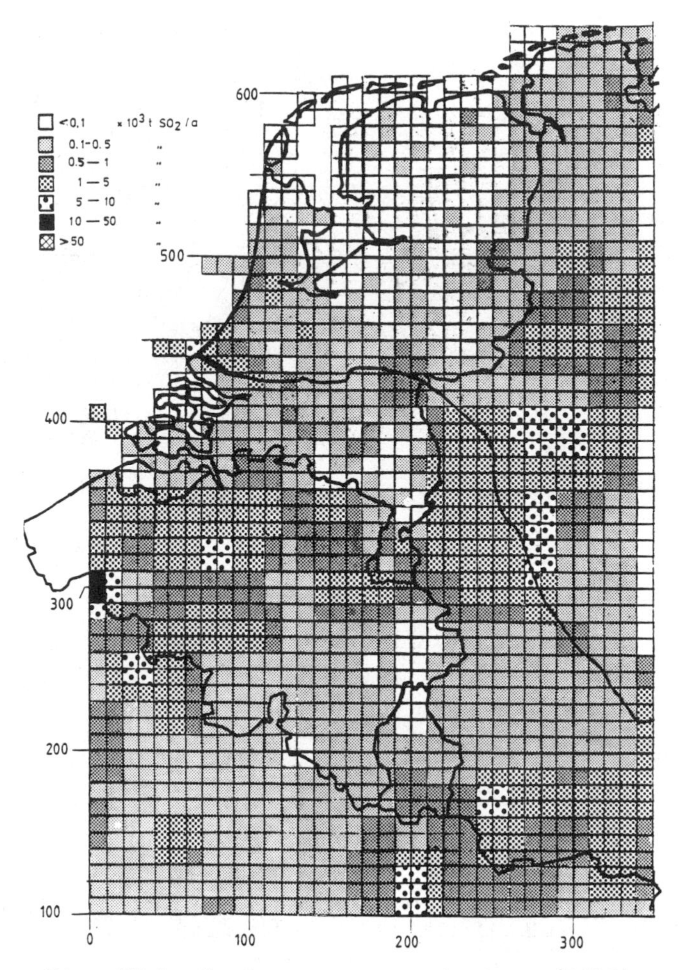

Fig. IV.2 Surface source emissions (1980).

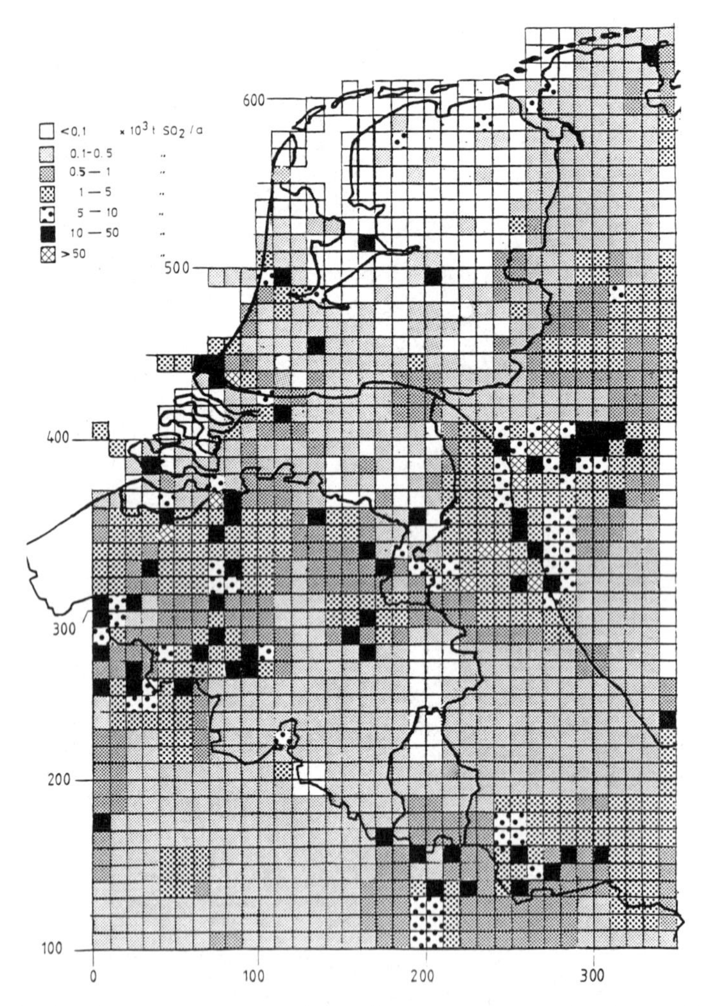

Fig. IV.3 Point source and surface source emissions (1980).

The primary metal industry in the Saar region has been aggregated into one point source with data from Metal Bull. Books Ltd (1978). Surface source emissions in the mid-west but outside the "Belastungsgebiete" were estimated with detailed statistical fuel-consumption data. All other surface source emissions were roughly estimated from BMI (1976). For space heating an emission factor of 0.765 kg SO_2/1000 inhabitants was used derived from the "Luftreinhaltepläne". Fuel consumption data were taken from IEA, OECD (1982), emission factors from UBA (1980).

Belgium, Luxemburg and France

Point sources were selected from Glückauf GmbH (1982), Metal Bull. Books Ltd. (1978) and Chem. Data Services (1977, 1978). Surface source emissions were allocated proportional to population density.

Netherlands

The emission file of the Project Emission Registration of the Ministry of Housing, Physical Planning and Environment was used.

Accuracy

It is hardly possible to make reliable statements concerning the accuracy of emission inventories. An overall accuracy of 20-30% might be reasonable for this inventory. The sum of all emissions is 4460 kt/a of which 70% is contributed by point sources, where power stations dominate. Of these, those burning more than one kind of fuel (notably in Belgium) are very difficult to assess. The accuracy of totals from point sources per country might be as follows:

Netherlands: 10%, Federal Republic of Germany 20%, Belgium and France: 30%. The accuracy of surface sources (25% space-heating) is certainly lower. Totals per country could vary from 10% in the Netherlands to 100% in France.

Remarks

a) The Paris region, lying near the lower left corner of the model area, has a roughly estimated emission of 450 kt/a.

b) It was formally arranged that this inventory would be verified by appropriate German authorities. This has yet to be carried out.

c) The emissions of the six large liquite-fired powerplants between Aachen and Köln are positively less than estimated in this inventory. Re-examination of these emissions resulted in 250 kt/a instead of 965 kt/a, Veldt (1983).

IV.b The meteorological data base and description of the episodes

One of the requirements was to use routinely available data only. Surface data were obtained from over 60 synoptic weather stations reporting one to three hourly. The data include the 10 m wind velocity, the surface pressure and temperature, precipitation (six hourly intervals) and cloud cover. In addition hereto, data from about 8 radio sondes were available and from which lapse rate and upper air wind velocity could be obtained. The release frequency of radio sondes was from 2 to 4 times per day.

The first episode, termed α, was a low-wind episode (21. Febr. 1979 00:00 - 22. Febr. 1979 24:00) where a strong high-pressure system (1045 mb) over the Soviet Union was slowly weakening. In the area chosen for the calculations there was a weak easterly wind. Strong inversions existed in this part of Europe. During the second day the wind direction was changing. The wind then came from the west. There was also a southwest component in the lower layers. An overview of the meteorological situation is shown in Figures IV.4 and IV.5 which are surface charts at 13.00 hours of the 21st and 22nd of February 1979.

The second episode, termed β, was a transport episode (13. May 1980 00:00 - 14. May 1980 24:00). Between a strong high pressure system over Scandinavia and a low pressure system above the Alps there was a continuous flow of continental air from the east. The wind velocities were between 5 and 15 ms^{-1}. In the upper air data there was a northeastern wind component. There were inversions in this episode, too. Figures IV.6 and IV.7 show the meteorological situation for these two days at 13.00 hours.

The last episode, termed γ, included a precipitation period (14. December 1981 00:00 - 15. December 1981 24:00). A depression system (975 mb) passed through the area from the northwest to the south-east. There was an easterly flow in the northern part and a westerly flow in the southern part of the area. Rain was falling at the warm front. An overview of the meteorological situation is given in Figures IV.8 and IV.9 for 13.00 hours.

IV.c Monitoring data base

Measured SO_2-concentrations (hourly averages) were available for comparison with the model results in the Netherlands and Belgium. The number of stations, method of measurement and interstation distances are shown in Table IV.2.

(text continued on page 221)

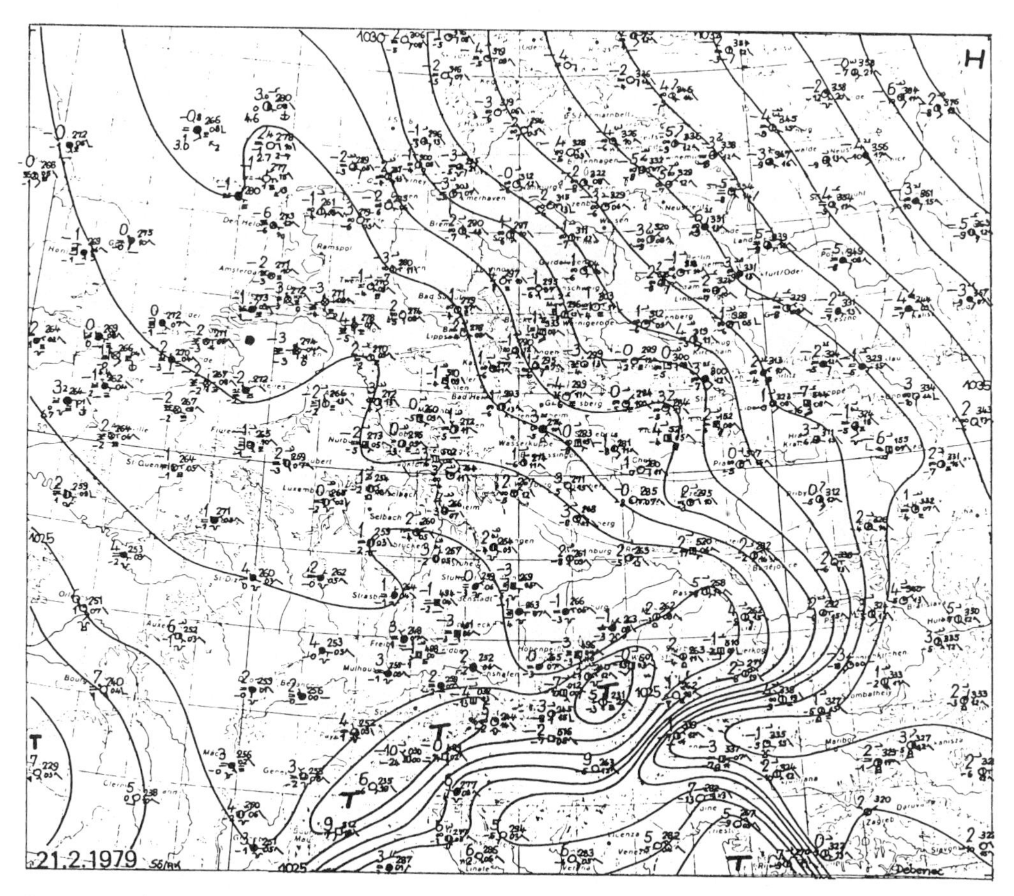

Fig. IV.4 Meteorological situation at 13.00 hrs. MEZ (21-2-1979).

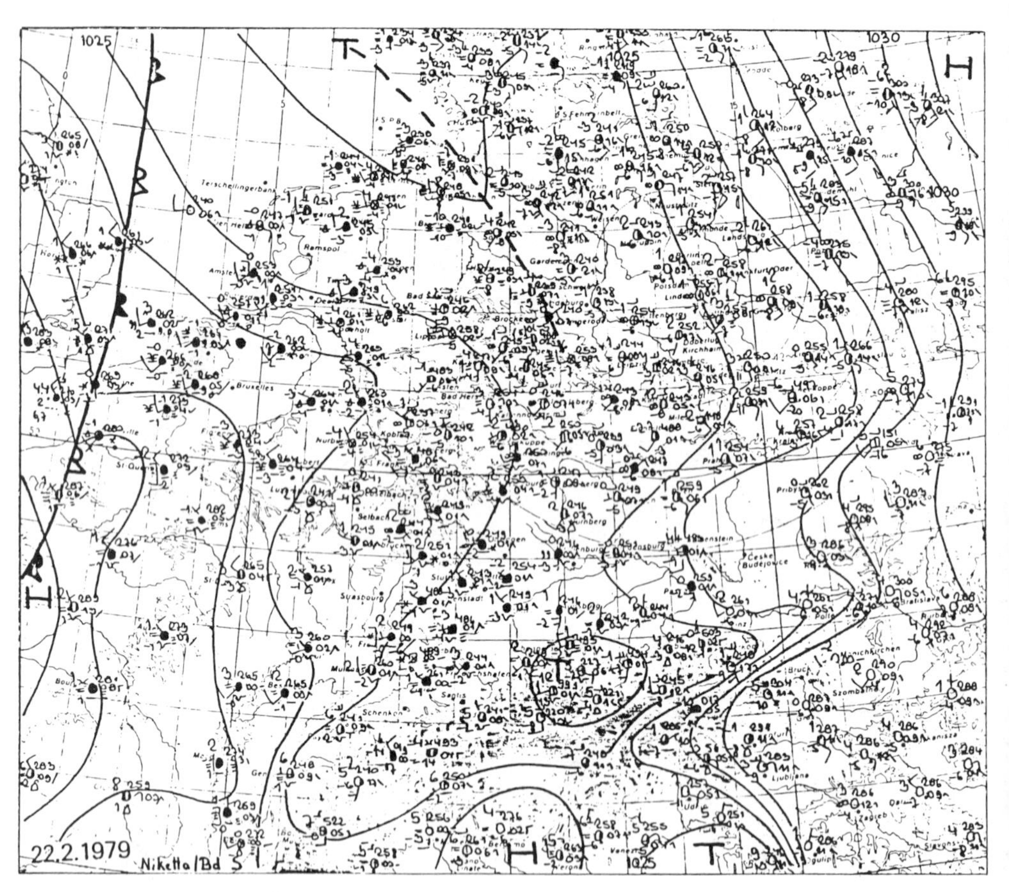

Fig. IV.5 Meteorological situation at 13.00 hrs MEZ (22-2-1979).

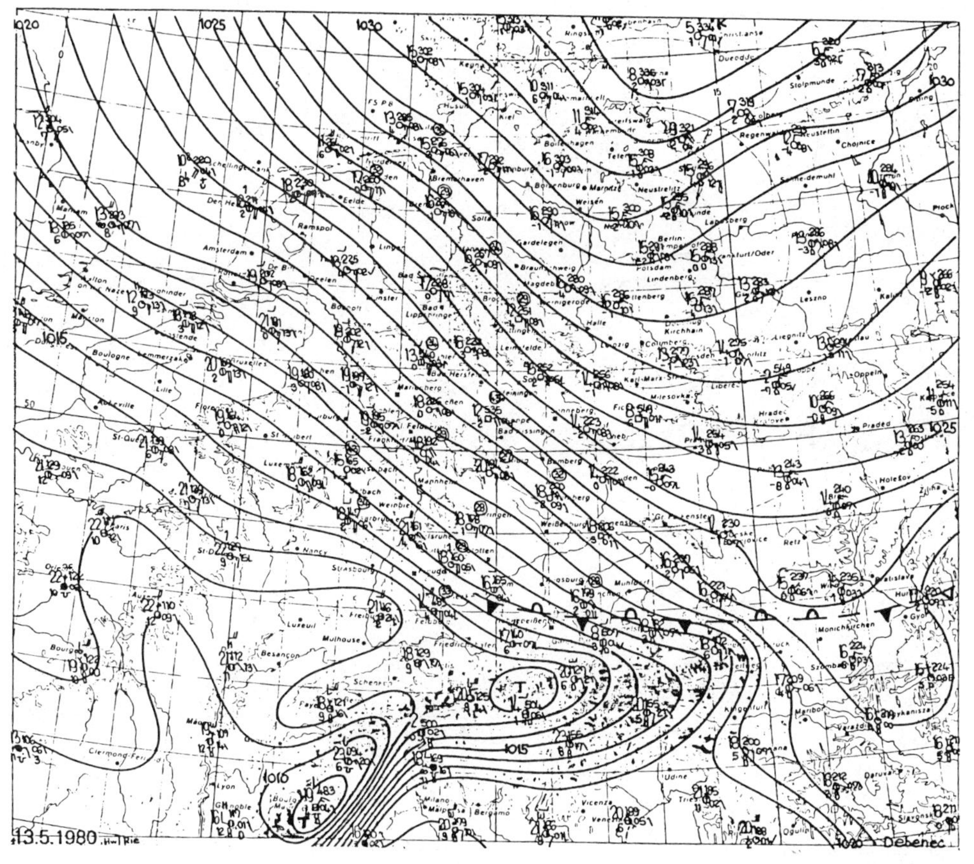

Fig. IV.6 Meteorological situation at 13.00 hrs. MEZ (13-5-1980).

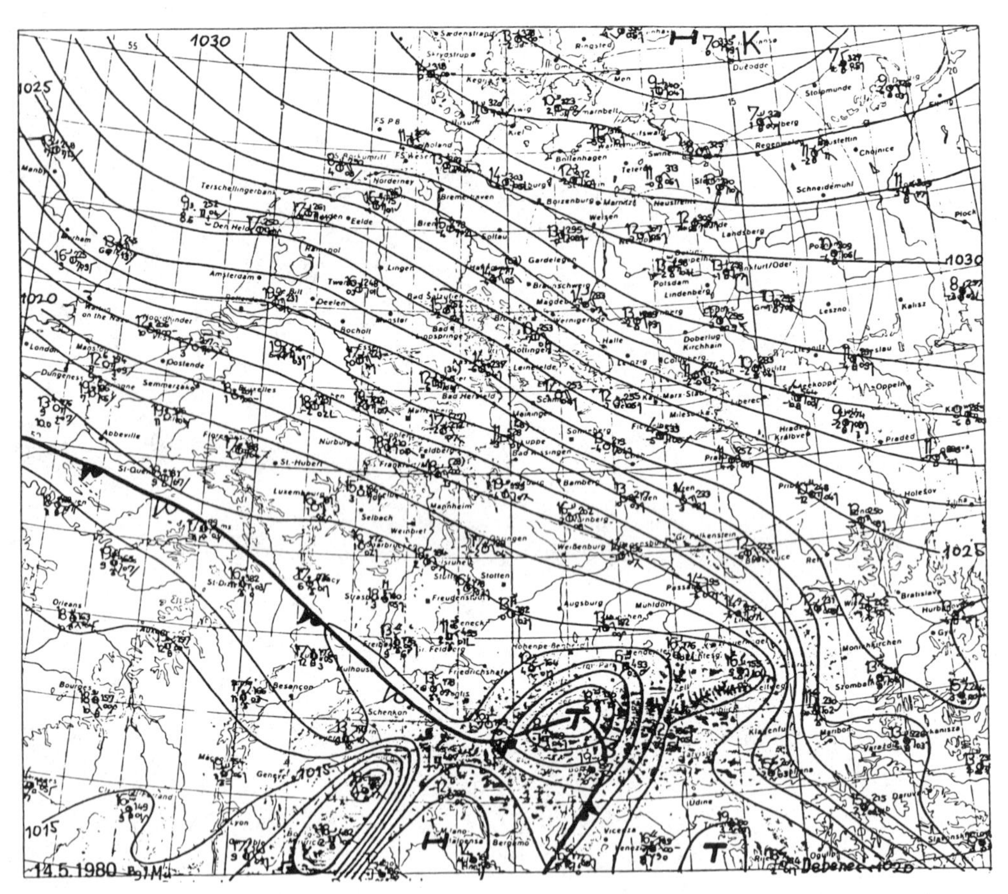

Fig. IV.7 Meteorological situation at 13.00 hrs. MEZ (14-5-1980).

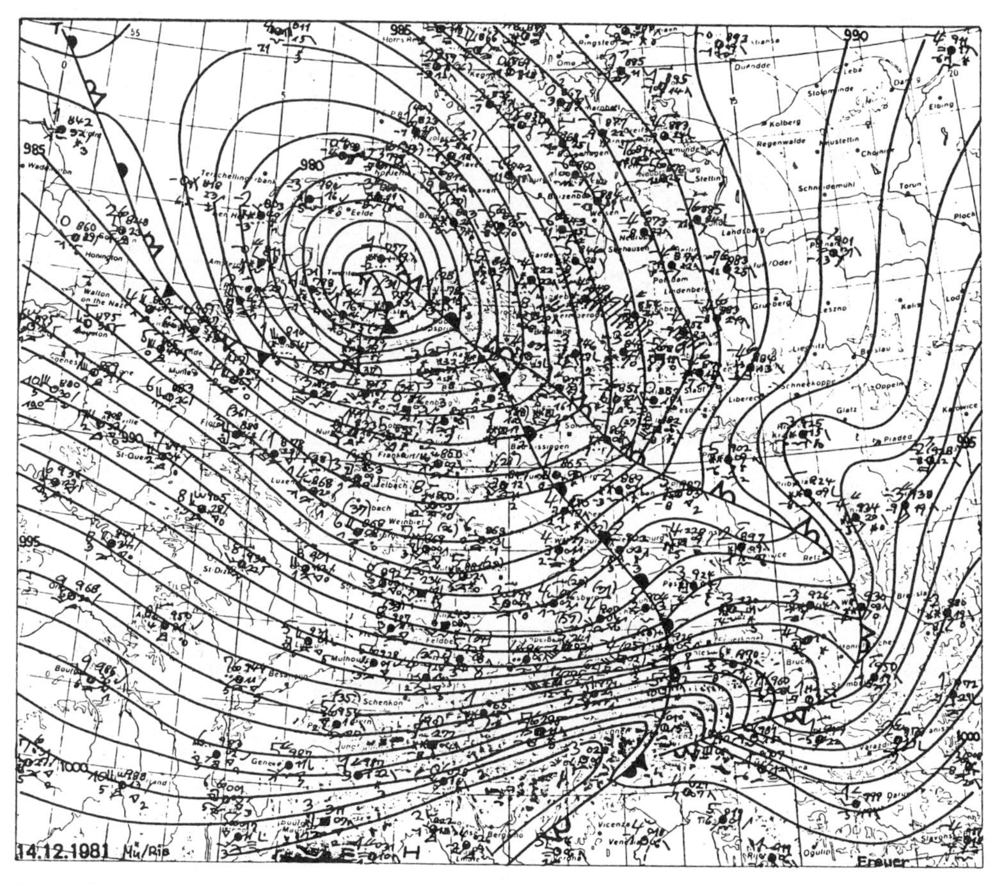

Fig. IV.8 Meteorological situation at 13.00 hrs. MEZ (14-12-1981).

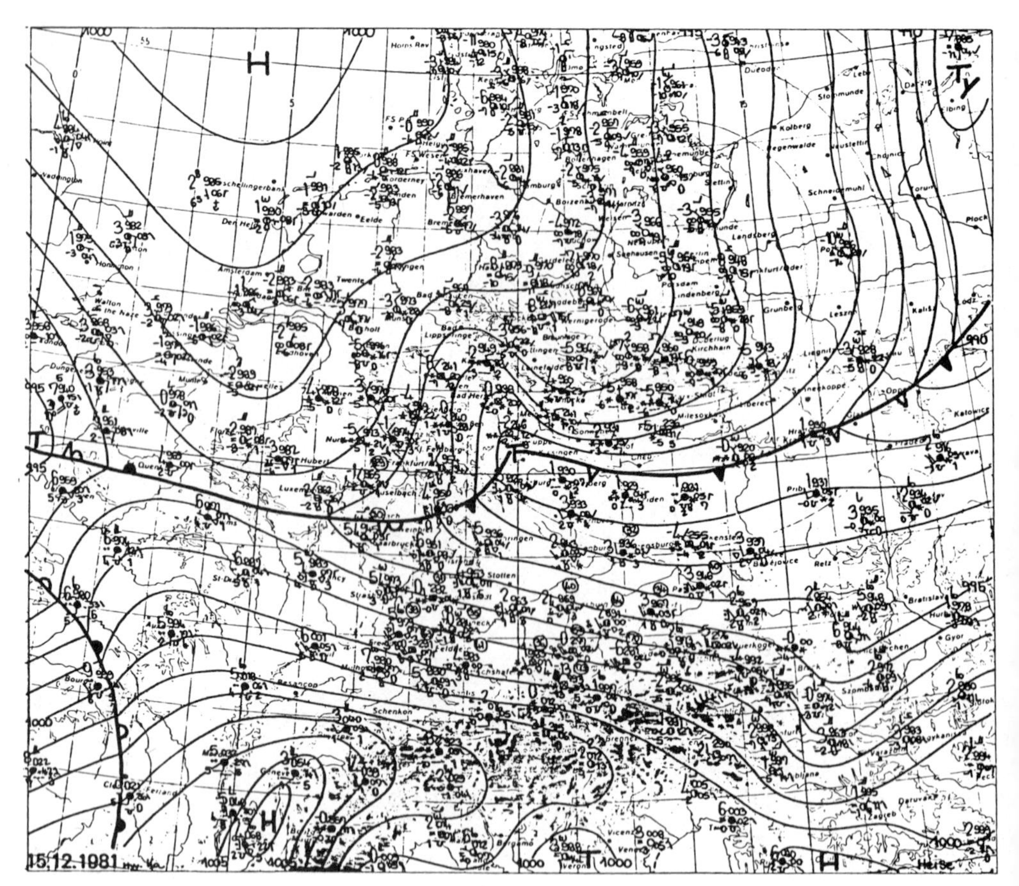

Fig. IV.9 Meteorological situation at 13.00 hrs MEZ (15-12-1981).

Table IV.2 SO_2 monitoring networks used in this study.

	Method	Number of stations	Interstation distances
The Netherlands	Philips PW 9700 (upgraded)	99	20 km
Belgium	Meloy	5 to 7	about 30 km

The monitoring stations were representative of a wider area. The measured concentrations are representative of the mesoscale concentration fields. Spatial interpolation is based on a negative exponential weighting scheme. Within the Netherlands, the concentration fields are reconstructed with an overall interpolation (standard) error of 20%.

For North-Rhine-Westfalia measurements were available for the α-episode at 3 places, for the β-episode at 12 places and for the γ-episode at 17 places. The interstation distances varied from 5 to 20 km. The measured values were interpolated according to the used grid.

V. ANALYSIS OF PHASE II AND III

General introduction

The way in which model results should be compared and evaluated has been a subject of discussion for a long time. One may apply the usual statistical tools such as correlation coefficients, r.m.s. differences etc., they too have been applied in this study. To look at these figures only, can be very misleading in some cases. Critical inspection and visual comparison of concentration fields has proven to be very useful in acquiring an impression of how the models work in a wide variety of (meteorological) situations.
In addition to statistical data, we have therefore decided to give descriptions of the time evolution of the concentration fields also.

In phase II there is one (hypothetical) continuous source of fixed strength and position. Differences in concentration patterns in this phase are due to the different processing of the meteorological fields, and to some extent also to different algorithms for transformation and deposition. It is obvious that large deviations in concentration fields occur when windfield directions differ by

10° only. This may be in cases where the wind direction varies strongly with height. All model results are extremely sensitive to vertical diffusion. The horizontal transport direction is also determined by the vertical extension of the plume to a large extent, and thus by vertical diffusion.

In phase III the total emission inventory has been applied. Here concentration fields are also strongly dependent on the spatial distributions of the common emission data base. Meteorological effects are therefore less obvious, and the results of the four models show in general a much better agreement than in phase II. In the next section we will give a survey of the three episodes. The first is a stagnant one, with parts of the area snow covered. The second is an advection dominated episode having quasi stationary characteristics. Finally the third episode is a highly complicated meteorological situation, where a depression moves over the centre of the area with accompanying precipitation.

V.a <u>Analysis of episode α (1979)</u>

The meteorological situation of the February episode is first determined by a strong high situated over eastern Europe. In the dispersion area one can find a weak south easterly wind and some stronger inversions.
During the second day, the situation is slowly changing to a more westerly flow caused by a new area of high pressure moving in from the west which removes the inversions and shifts the concentrations to the east. In general there are low windspeeds in the period. During the second day especially, when the comparison between the models is made, wind speed is extremely low from 0.00 hours to 4.00 hours, weakly north west from 4.00 to 13.00 hours and then backing to west-south-west. Given the low wind conditions, pollution transport is dominated by vertical and horizontal dispersion rather than by advection.

<u>Analysis of Phase II:</u>

The phase II computation generally display differences in vertical mixing (e.g. fumigation), advection and horizontal dispersion as simulated by the respective models. Differences in vertical stability and mixing height profiles result in large discrepancies from 0.00 hours to 10.00 hours as illustrated by the concentration fields at 7.00 hours (Figure V.1). Horizontal dispersion and advection treatment at 22.00 hours finally leads to different geographical positions of the core of the plume (Figure V.4).

The treatment of vertical stability and mixing differs strongly between the modellers. As illustrated by Figure II.5, KNMI and TNO (SAI) simulate a low mixing height in the early morning hours, increasing from 9.00 hours to the end of the afternoon

(TNO(SAI)) or till 15.00 hours (KNMI). The RIV-model is run with a constant mixing height of 250 m. In the THD-model no explicit mixing height is defined; vertical mixing is determined by the K_z-field. Turbulent diffusion is low during the early morning hours, higher during the day and decreasing in the evening.

Due to the low mixing height at KNMI and TNO (SAI) effective emission (225 m) takes place above the ground based layer(s) and ground concentration remain low until fumigation starts at 9.00 hours. THD and RIV models emit below mixing height showing high ground level concentration (see Figures V.1, V.2). At 10.00 hours (Figure V.2) KNMI and TNO (SAI) ground level concentrations are increased by fumigation. THD and RIV display high concentration near the source while KNMI and TNO (SAI) show lower maximum concentration. In TNO (SAI) the low mixing height is restored at 16.00 hours and the 225 m source emission is cut off from the surface layer. Consequently the ground level plume is decoupled from the source as illustrated by Figure (V.4).

During the first half of the day advection is weak; nevertheless, however, the positions of the plumes display a discrepancy for the TNO (SAI) model; at 13.00 hours (Figure V.3) this model shows the major extension to the north, while the other models display advection to the south. This results in lower correlation coefficients for TNO (SAI) with the other models (Figure V.6).

From 13.00 hours on correlation of TNO (SAI) with RIV increases in contrast with THD and KNMI. This has arisen through differences in advection during the second half of the day. RIV and TNO (SAI) maintain low mixing heights after 16.00 hours. Advection in the ground layers is consequently dominated by the surface wind field which, as could be expected, is more backed to the SW than the veered westerly winds of higher levels as effectively used by THD and KNMI. As illustrated by Figure V.4, this results in plume core axis orientations to NE for RIV and TNO (SAI) and E to SE for THD and KNMI. The KNMI model shows a truncation effect from 19.00 hours on; towards the East concentrations decrease to zero thus resulting in strong concentration gradients.

At 22.00 hours (Figure V.4), concentrations at the eastern part of the field are consequently zero. This is probably due to the numerical advection scheme.

The strongest horizontal dispersion is simulated by the RIV model, this results in low gradients of the concentration field at 13.00 hours (Figure V.3) and 22.00 hours (Figure V.4). The TNO (SAI) model is characterized by box shape plume patterns having low gradients in the plume centre and strong gradients near the plume edges.

(text continued on page 229)

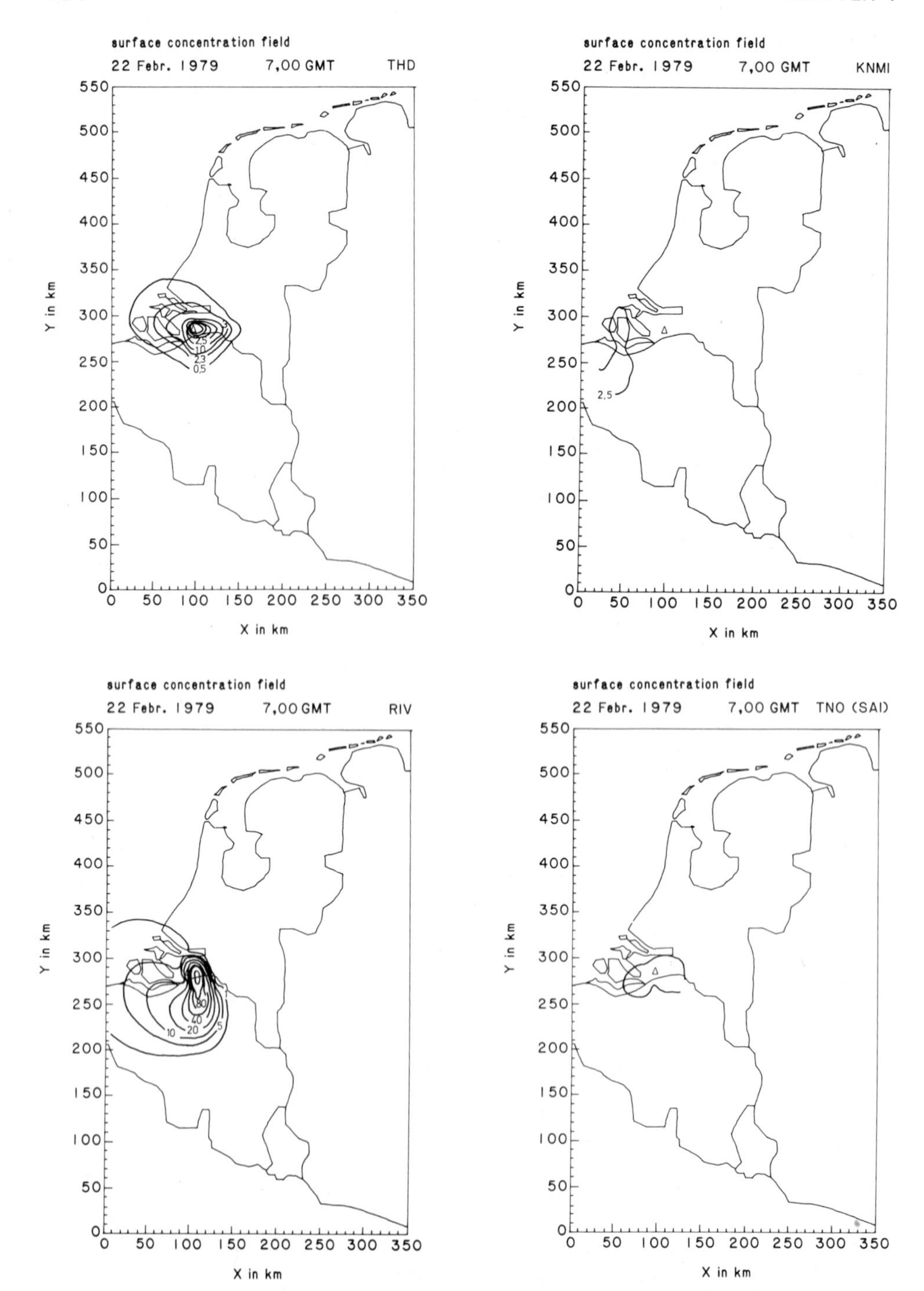

Fig. V.1 Surface concentration fields for the different models in episode α (phase II) at 7.00 GMT; concentration in μg/m³.

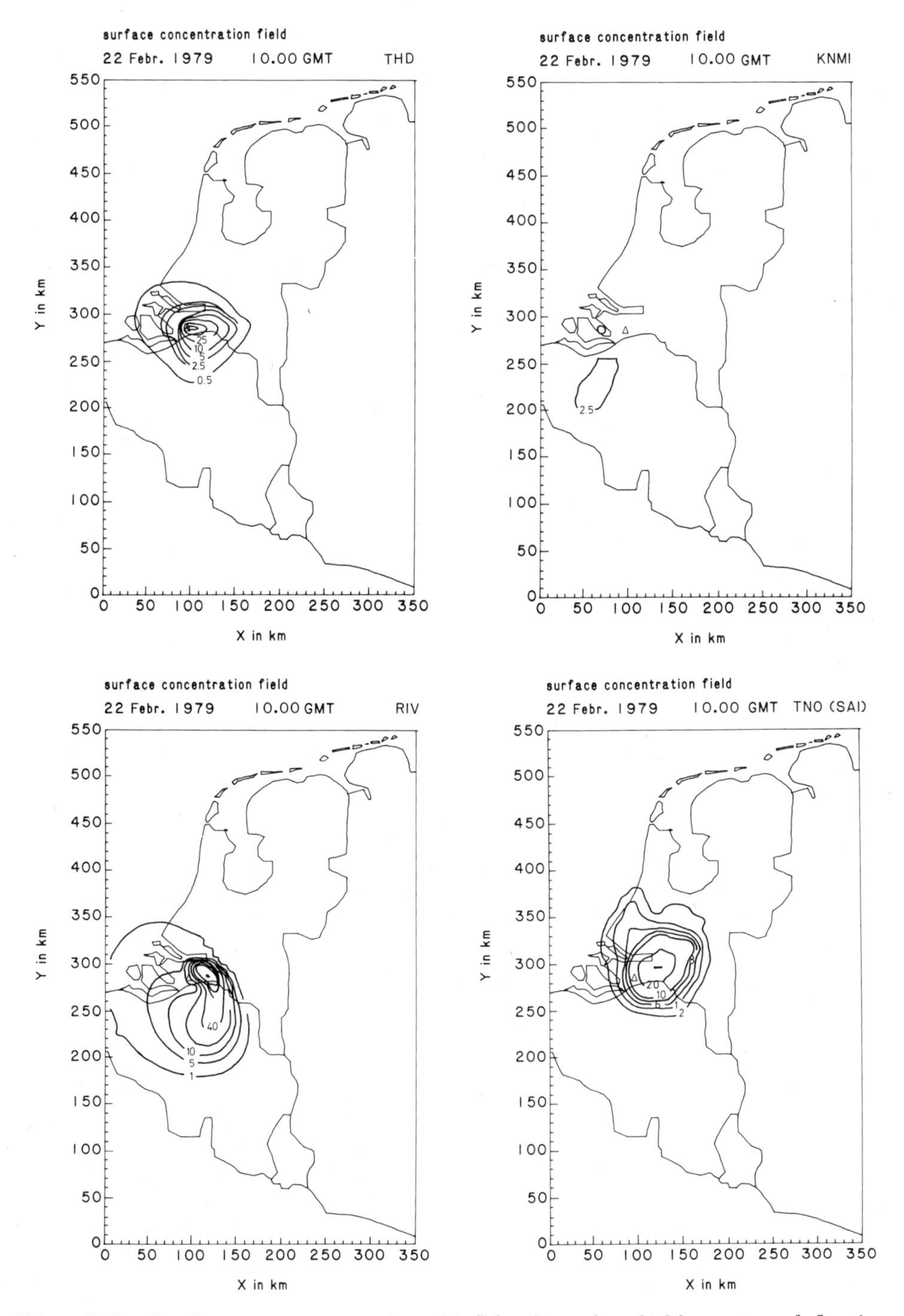

Fig. V.2 Surface concentration fields for the different models in episode α (phase II) at 10.00 GMT; concentration in μg/m³.

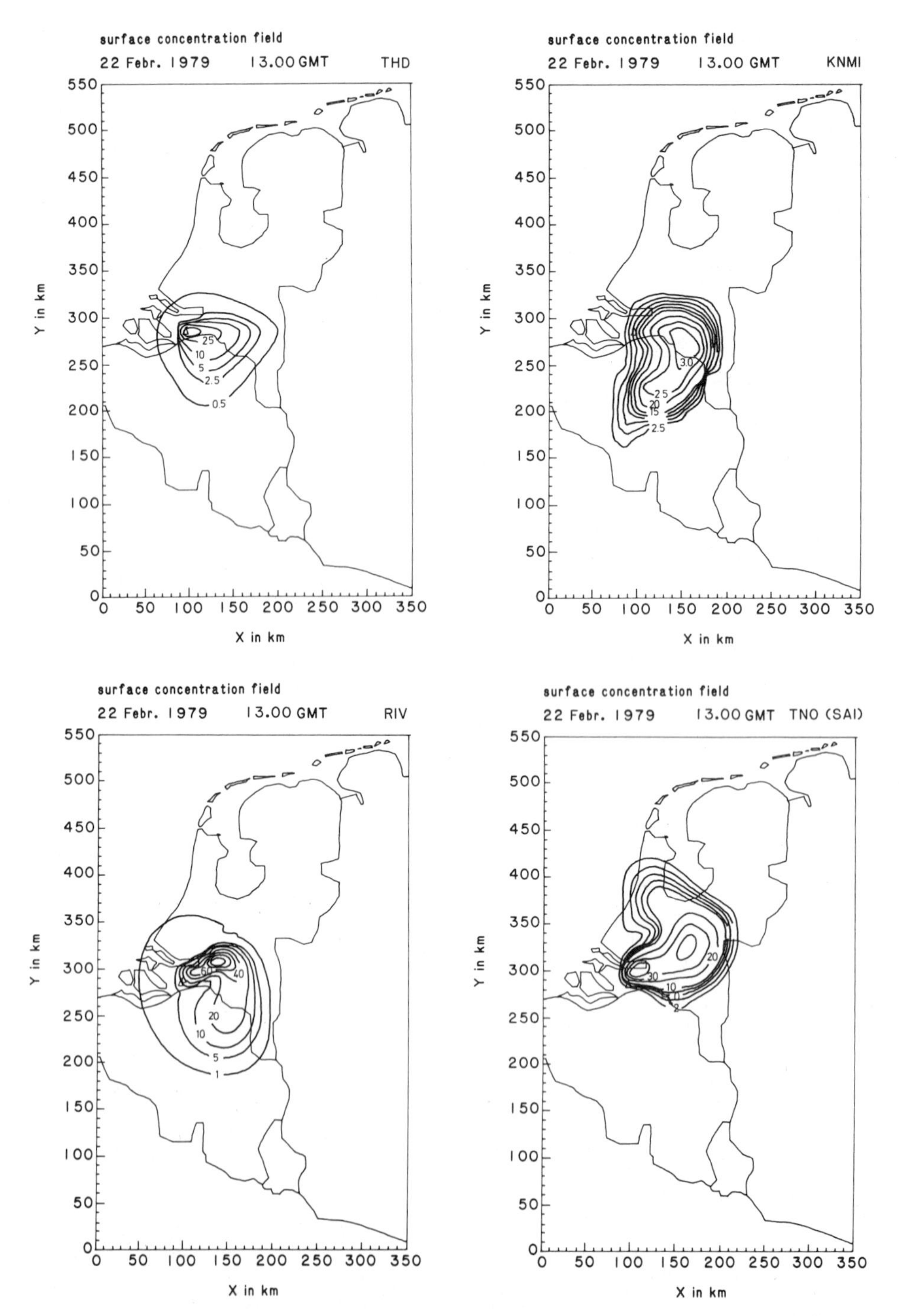

Fig. V.3 Surface concentration fields for the different models in episode α (phase II) at 13.00 GMT; concentration in $\mu g/m^3$.

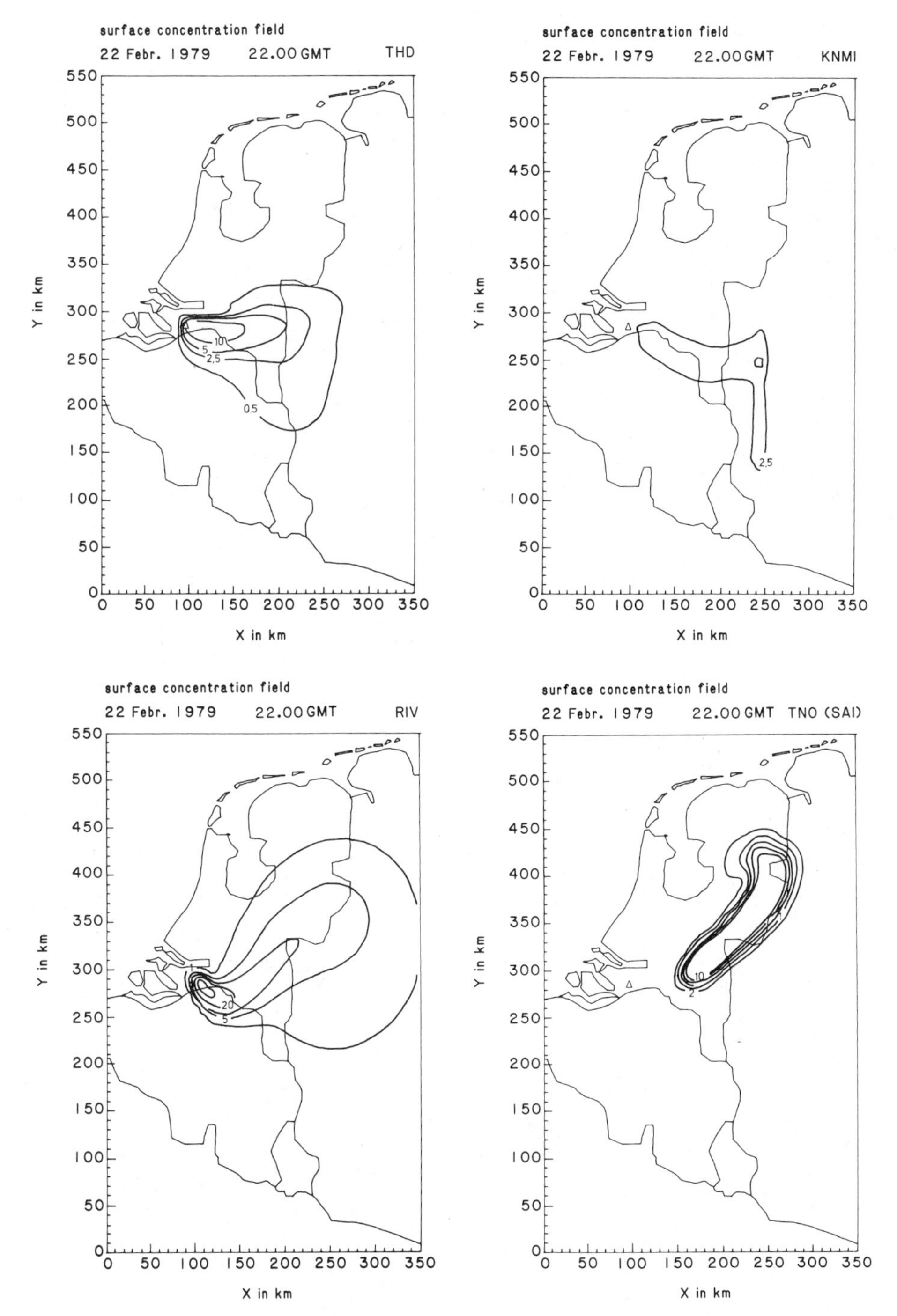

Fig. V.4 Surface concentration fields for the different models in episode α (phase II) at 22.00 GMT; concentration in μg/m³.

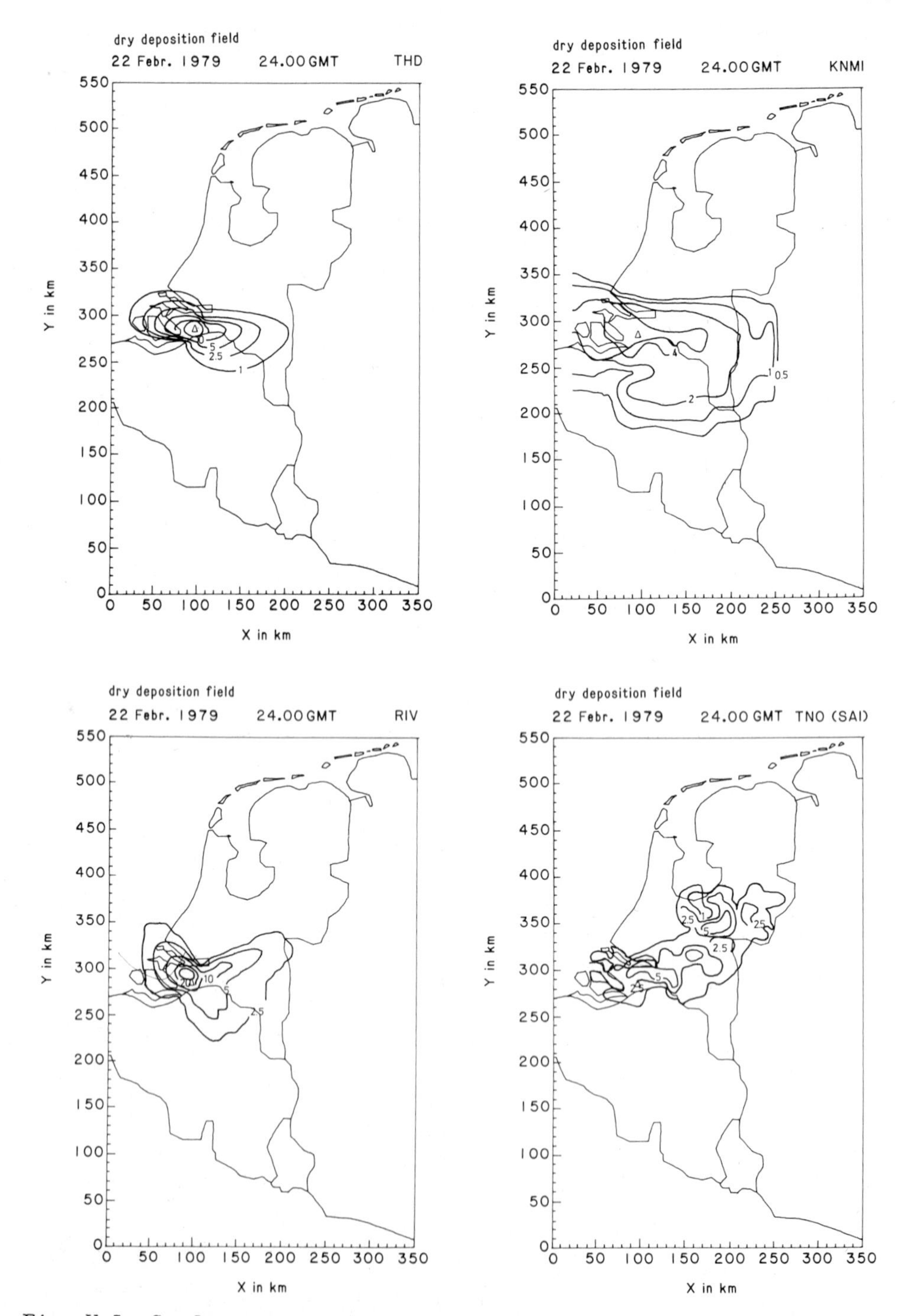

Fig. V.5 Surface concentration fields for the different models in episode α (phase II) at 24.00 GMT; concentration in μg/m³.

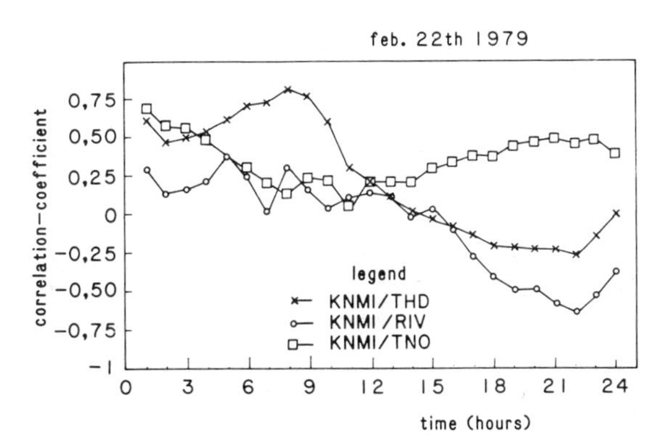

Fig. V.6 Correlation-coefficient TNO(SAI)/models; phase II and episode α.

KNMI and TNO (SAI) models show also decoupling of the ground concentration maxima (plumes) from the source as illustrated by the fields of 7.00 and 22.00 hours (Figure V.1 and V.4). This cutting off arises from a realistic modelling of the vertical stability. A slight decoupling is additionally observed in the KNMI model and which is the result of allocating the initial concentrations to downwind grid elements in order to avoid high spatial frequences in the pseudo-spectral advection scheme. This effect was mentioned earlier herein with respect to the phase I results.

The mutual correlations between the deposition fields for THD, KNMI and RIV are relatively high in comparison with the TNO (SAI) correlations (Figure V.5). This can be explained by the more northerly directed advection in the TNO model during the first half of the day. As a result of the correspondence in advection between TNO (SAI) and RIV towards the end of the day, deposition fields of TNO (SAI) and RIV show no deposition in the NE direction.

The deposition pattern of KNMI and especially that of TNO (SAI) reflect the more subtile treatment of dry deposition as a function of land use.

Highest deposition rates are computed by RIV; the other models compute about the same depositions. This is explained by the presence of snow cover which was not accounted for in the modified RIV model; in the latter, deposition velocity and surface layer concentration gradient had to be derived from Pasquill stability classes in contrast to the original model in which the parameters were derived from radiation measurements and albedo.

Analysis of phase III

In relation to the weak advection during the first half of the day and the more stationary W-SW transport during the second half of the day, the correlations between models and measurements increase towards the end of the day (Figure V.13). As illustrated by the measured concentration fields at 7.00 hours, a puff of concentrations between 500 and 1000 μg/m³ SO_2 is built up in the Rotterdam-Rijnmond area during the night and firstly transported in ESE-direction and after 13.00 hours advected in ENE-direction.

None of the models simulates this "puff" adequately, this explains the low correlations with measurements in the beginning of the day. The mutual correlations between the models are high during the first half of the day. During the second half when the pollution is advected to the NE correlations between the models decrease and correlations with measurements increase (Figure V.12, V.13). The decrease in correlations between the models cannot be explained from the phase II results where higher correlations were found between TNO (SAI) and RIV due to the correspondence in advection direction. In the phase III results (Figure V.10), the decoupling of sources from ground level in the Ruhr area, as a consequence of the low nocturnal

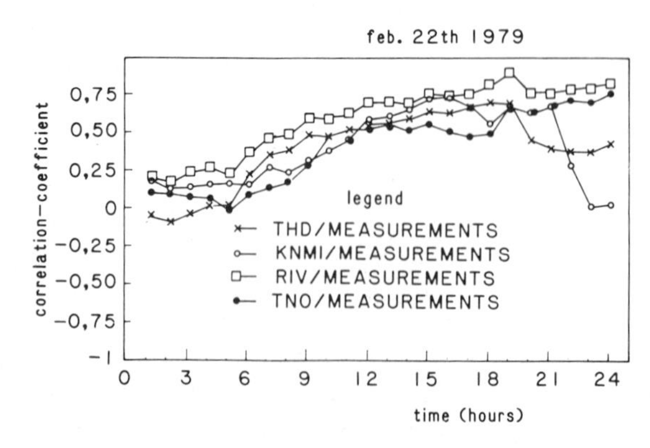

Fig. V.13 Correlation-coefficient of models/measurements (episode α Phase III).

mixing height in the TNO (SAI) model, leads to lower correlations with the other models. Apparently source and mixing height effects dominate the advection effects.

The RIV model shows an instantaneous increase of concentration by a factor of two at 10.00 hours, this results in large discrepancies with the others (Figure V.12a). This is due to the abrupt change in the surface layer gradient which, as stated above, arises from changing from night to day conditions in the simplified Pasquill-scheme.
Thus during daytime high concentrations, especially in the Ruhr area, are computed by the RIV model as a result from the low (250 m) and constant mixing height.

As could be expected, from the phase II results highest depositions are computed by the RIV model.
Apart from the locally high values in the Ruhr area for the RIV model and the land use effects in TNO (SAI), the correspondence with respect to both shape and absolute level between the deposition fields is rather good (Figure V.11).

V.b <u>Analysis of episode β (1980)</u>

During this episode a strong easterly wind dominated, the mixing height was over 900 m all the time and all sources were well within the mixed layer.

<u>Analysis of phase II:</u>

Examining the output data of the models, we see a large similarity in the pattern (Figures V.14, V.15). After the night period the values of the concentrations at the plume axis agree within a factor of two. The advection has been well described by all models. The most striking difference in the patterns are the low values of the TNO (SAI) model output during the night time. A lower mixing layer height has been used, which prevented the downward flux from the 225 m high stack situated in the Ruhr area. This fact is also illustrated by the time series of the root mean square error and the relation of mean values of the surface data of TNO against the other models (Figure V.17).
We see a discontinuity at 6.00 GMT, at the moment the TNO model output displays a simulation of the fumigation process.

The RIV model output shows a faster decrease with distance of the surface concentration in the plume axis than the others. This can be explained by the relatively high surface values near the source. Deposition near the source is therefore larger.

(text continued on page 242)

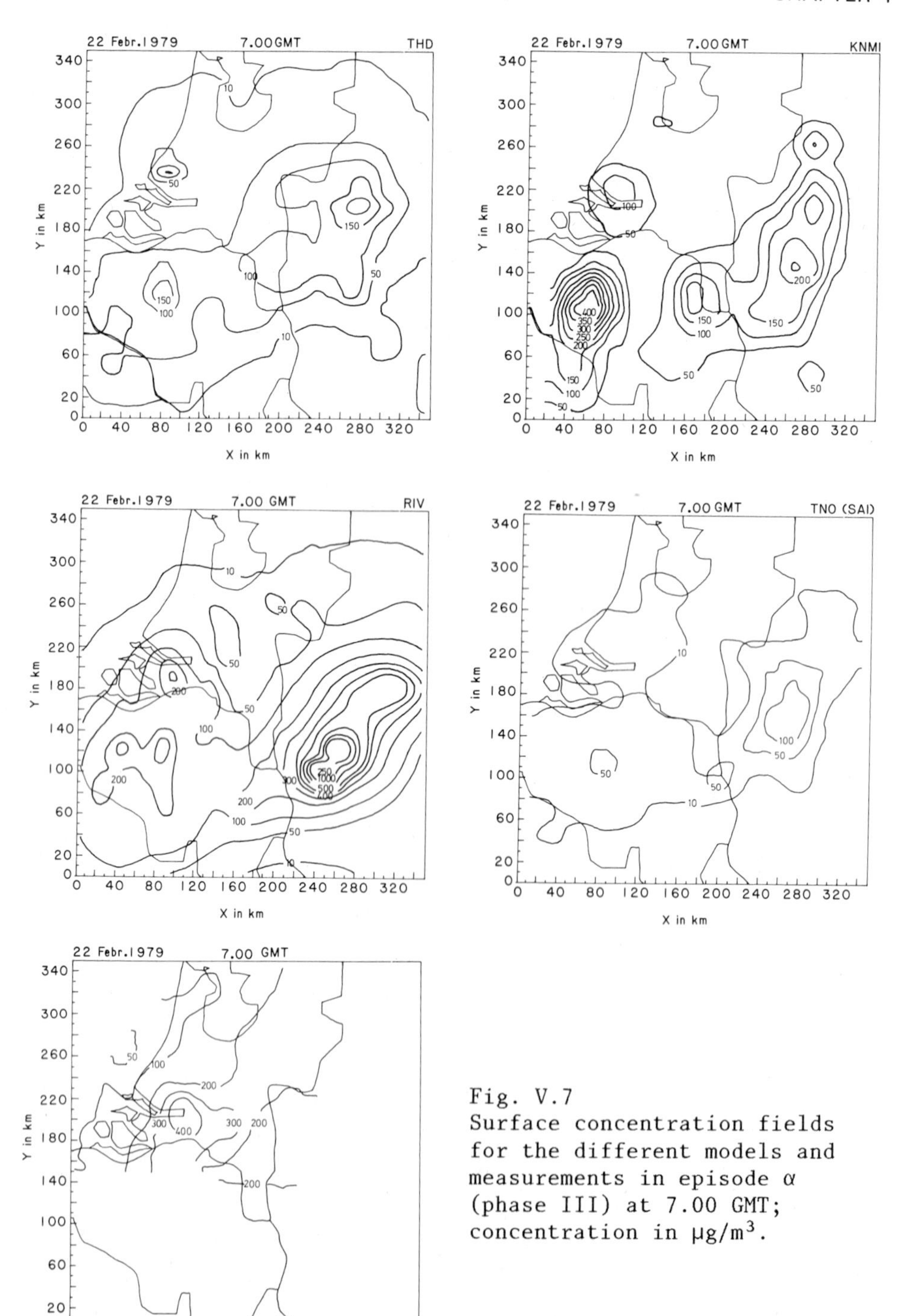

Fig. V.7
Surface concentration fields for the different models and measurements in episode α (phase III) at 7.00 GMT; concentration in μg/m³.

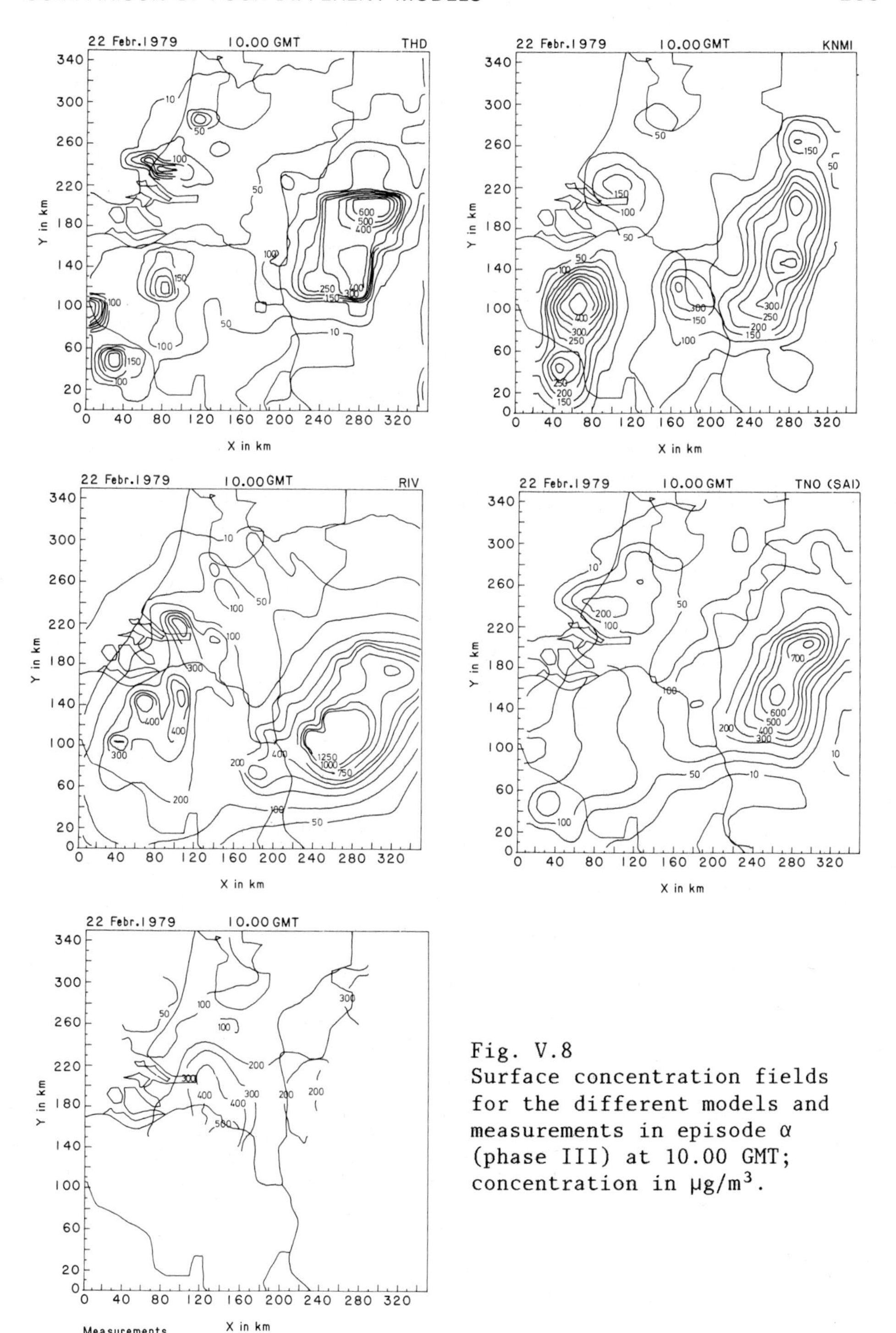

Fig. V.8
Surface concentration fields for the different models and measurements in episode α (phase III) at 10.00 GMT; concentration in μg/m^3.

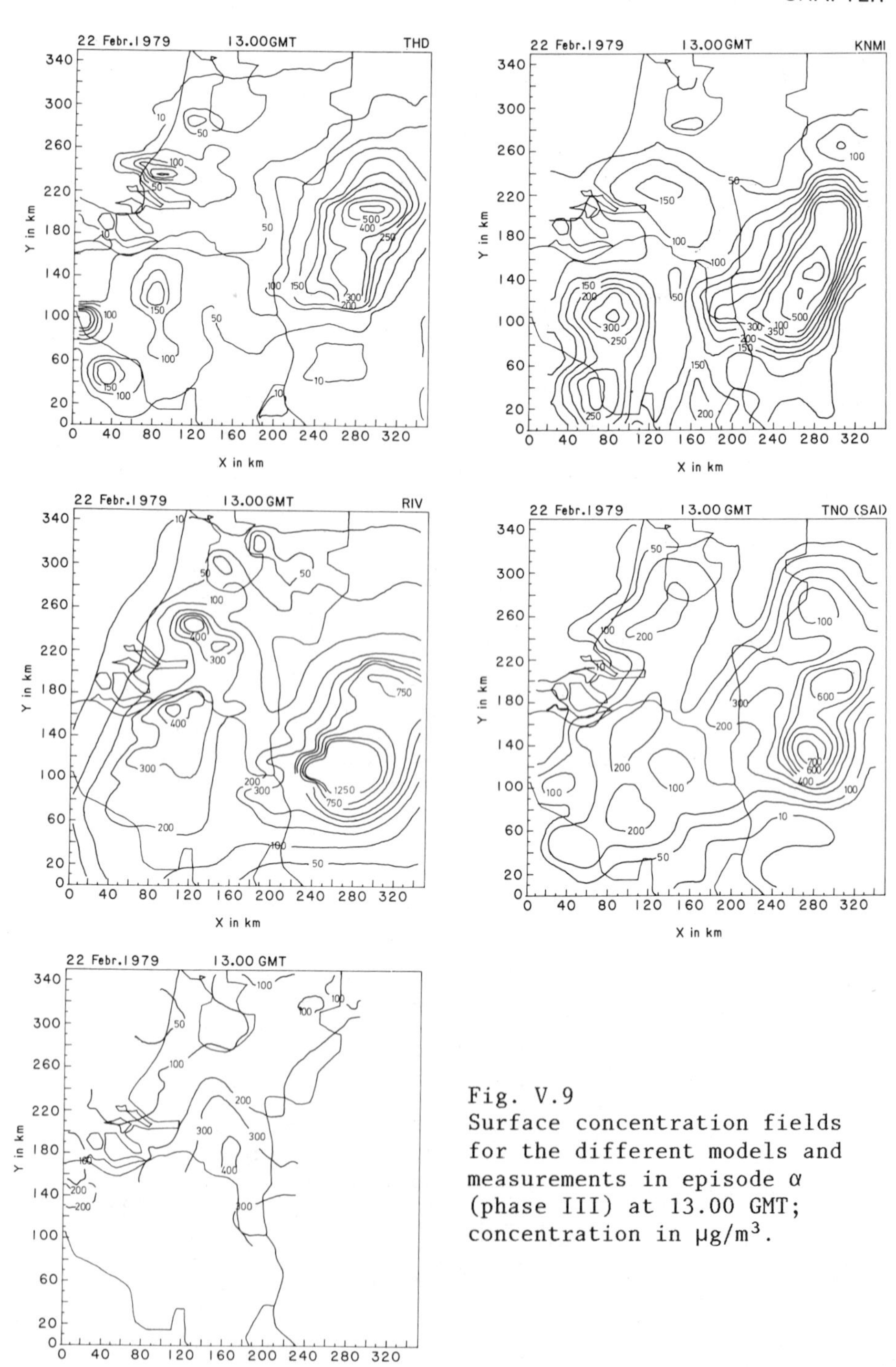

Fig. V.9
Surface concentration fields for the different models and measurements in episode α (phase III) at 13.00 GMT; concentration in $\mu g/m^3$.

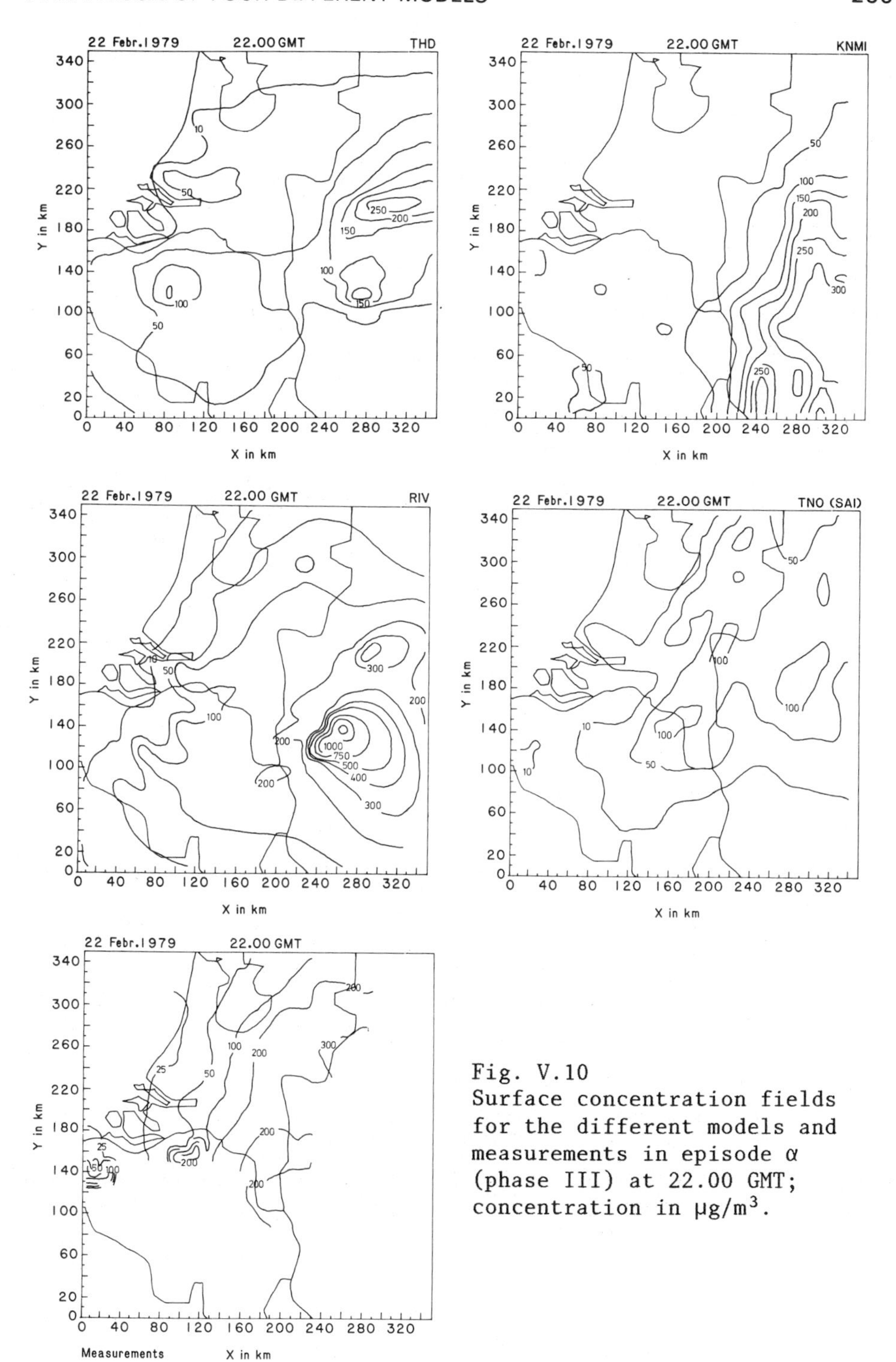

Fig. V.10
Surface concentration fields for the different models and measurements in episode α (phase III) at 22.00 GMT; concentration in μg/m³.

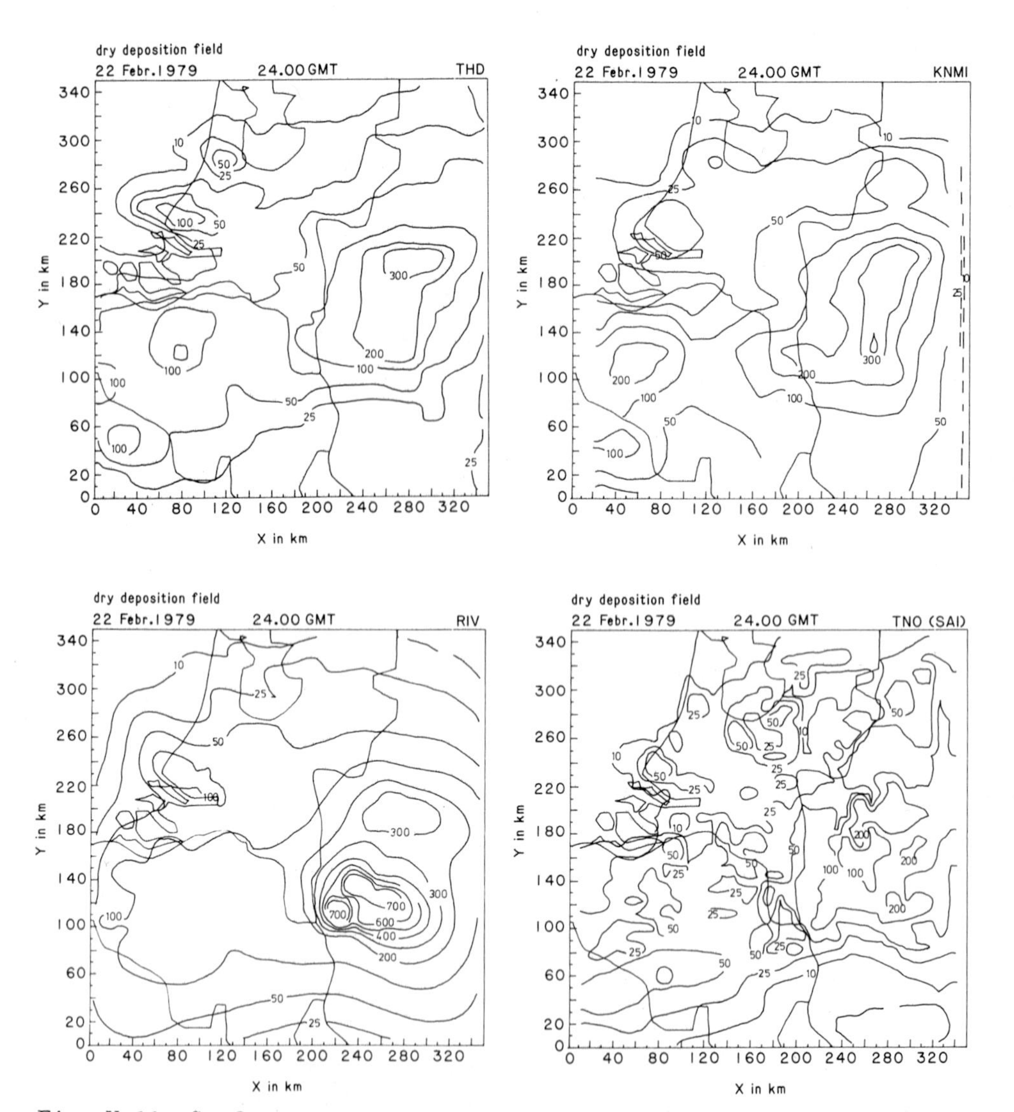

Fig. V.11 Surface concentration fields for the different models and measurements in episode α (phase III) at 24.00 GMT; deposition in mg/m^2.

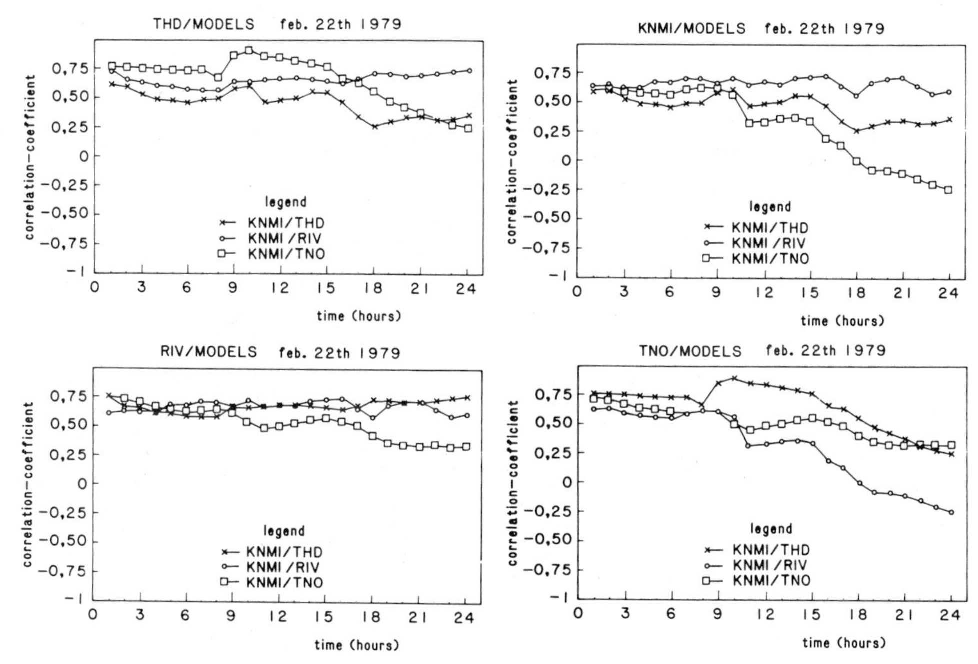

Fig. V.12 Correlation-coefficient between models (phase III; episode α).

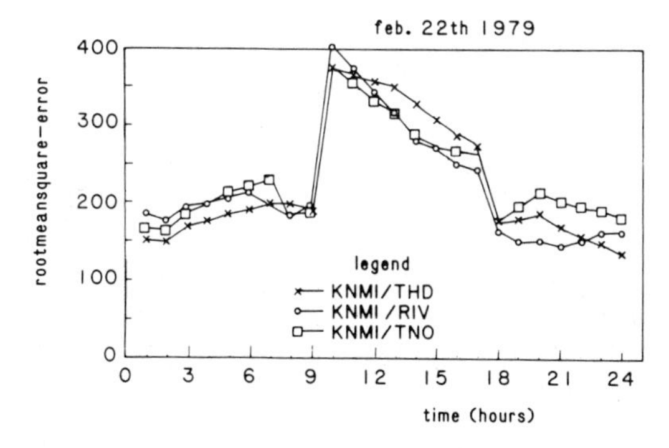

Fig. V.12a Root-mean-square-error RIV/models (phase III, episode α) in $\mu g/m^3$.

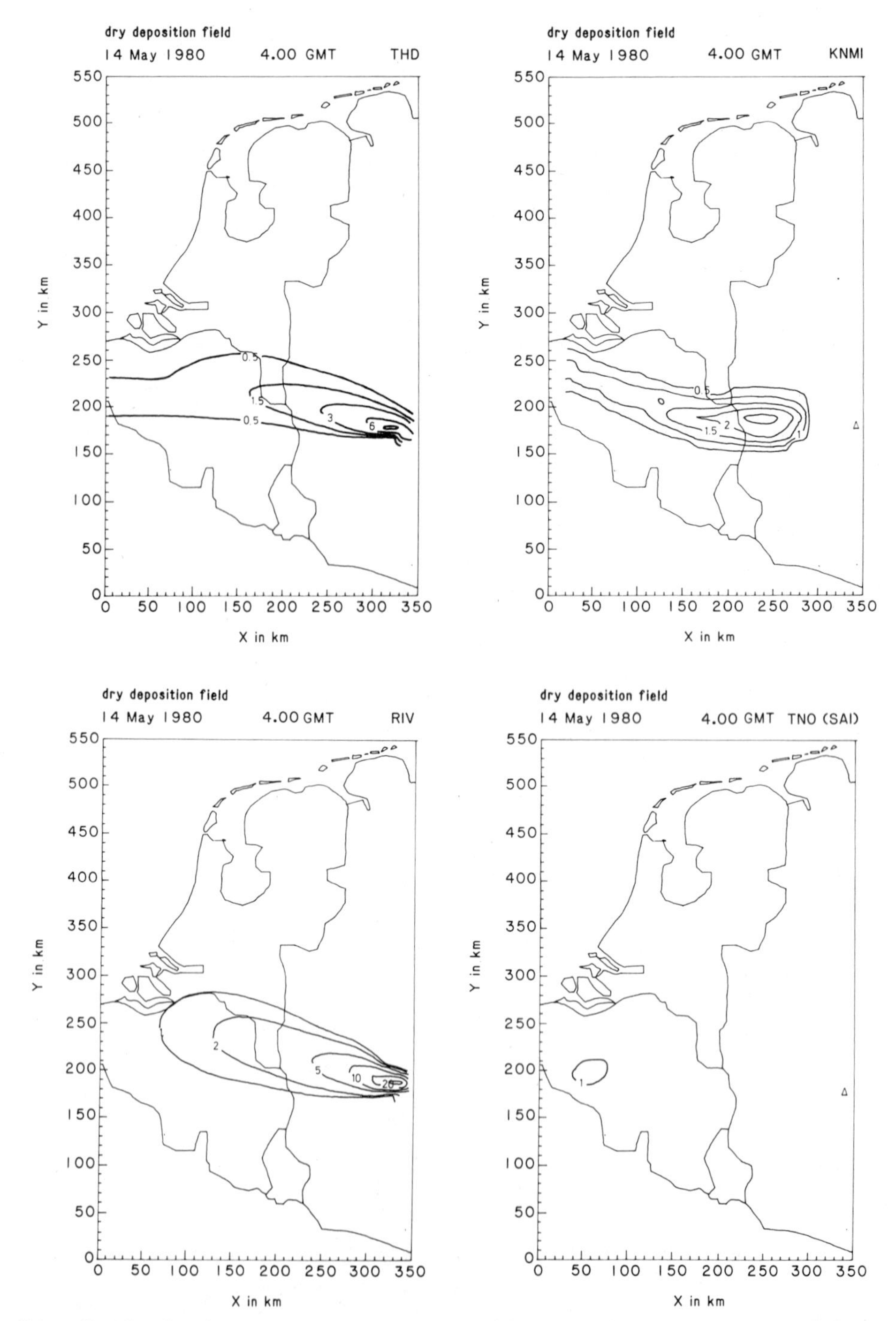

Fig. V.14 Surface concentration fields for the different models in episode β (phase II) at 4.00 GMT; concentration in μg/m³.

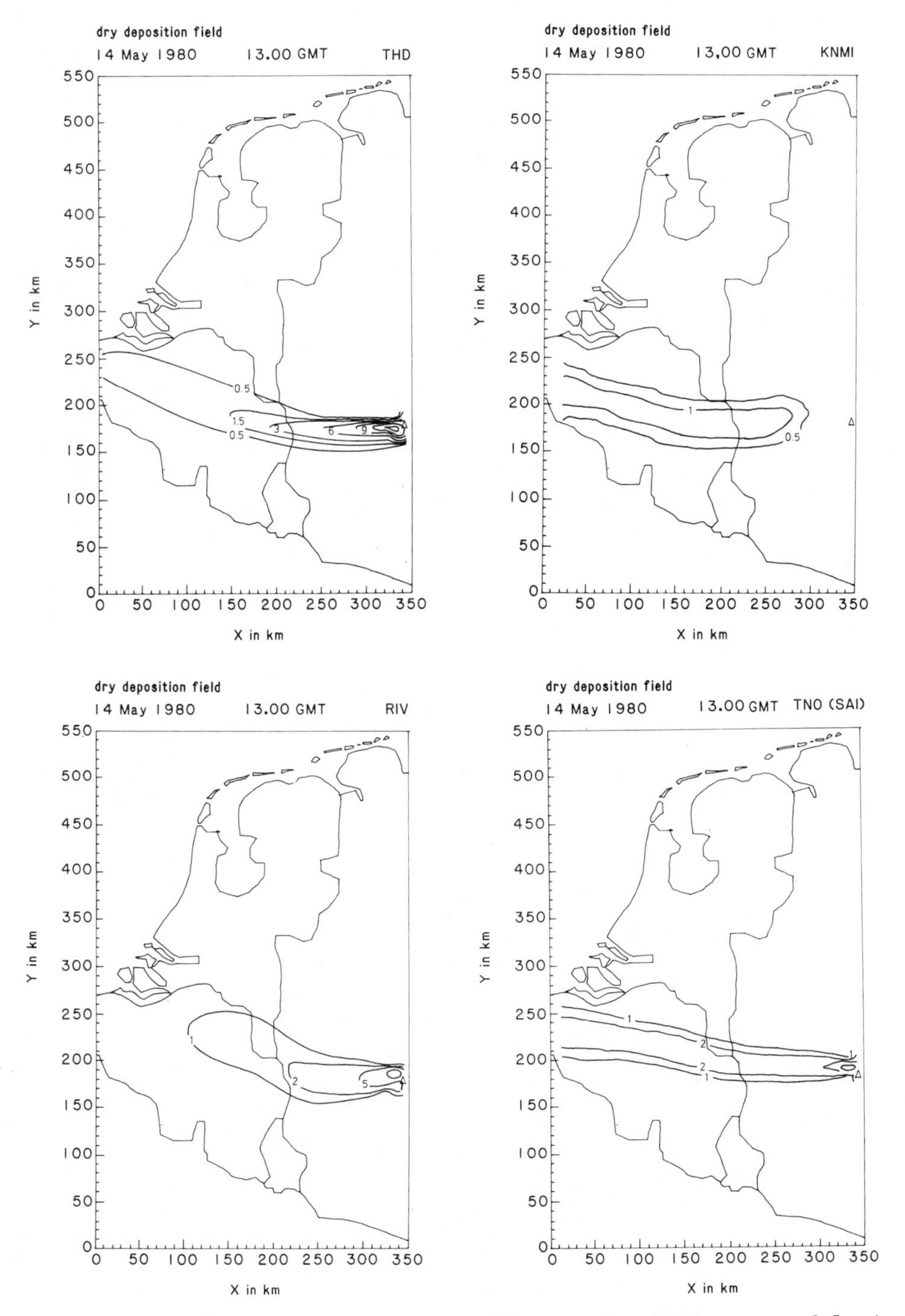

Fig. V.15 Surface concentration fields for the different models in episode β (phase II) at 13.00 GMT; concentration in $\mu g/m^3$.

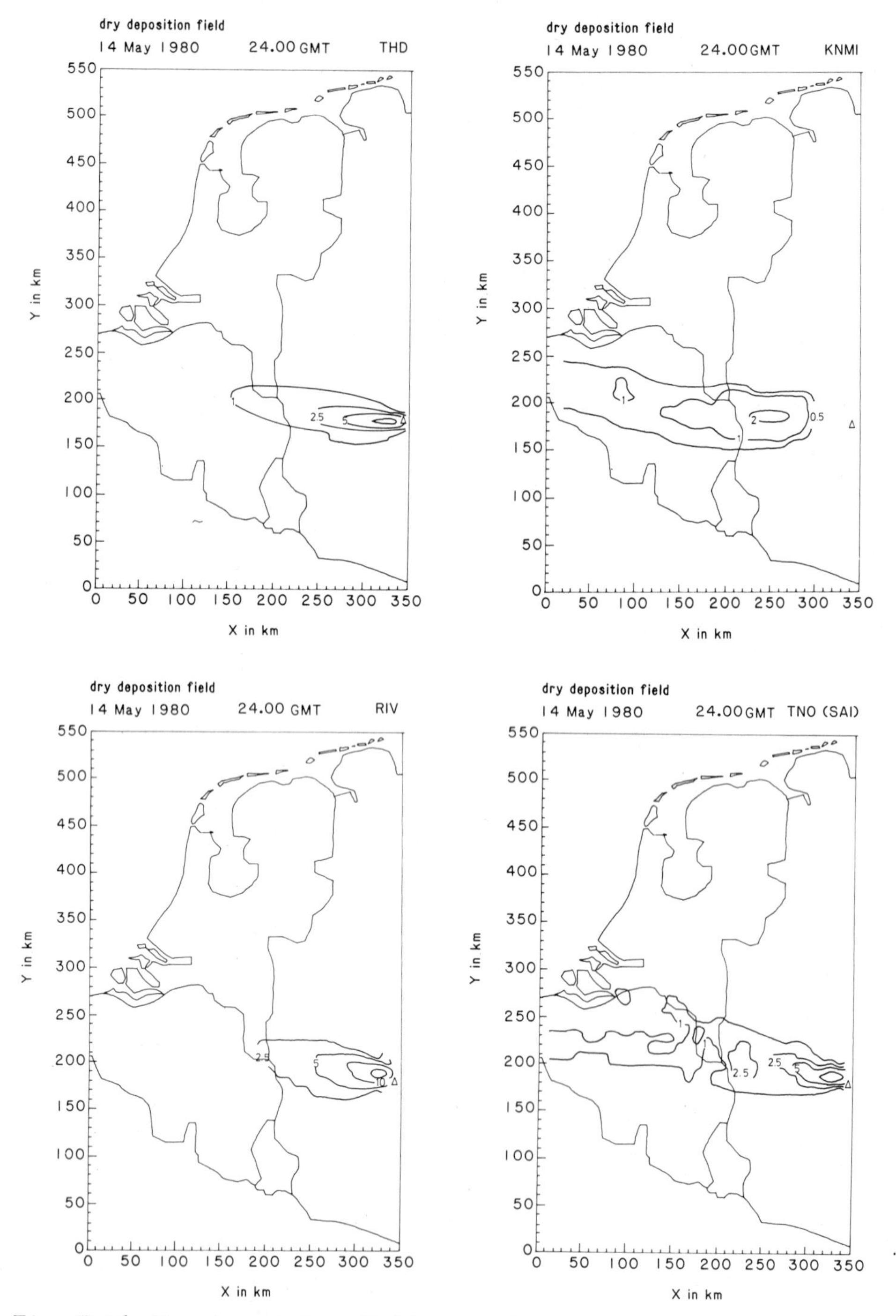

Fig. V.16 Dry deposition fields for the different models in episode β (phase II) at 24.00 GMT; deposition in mg/m^2.

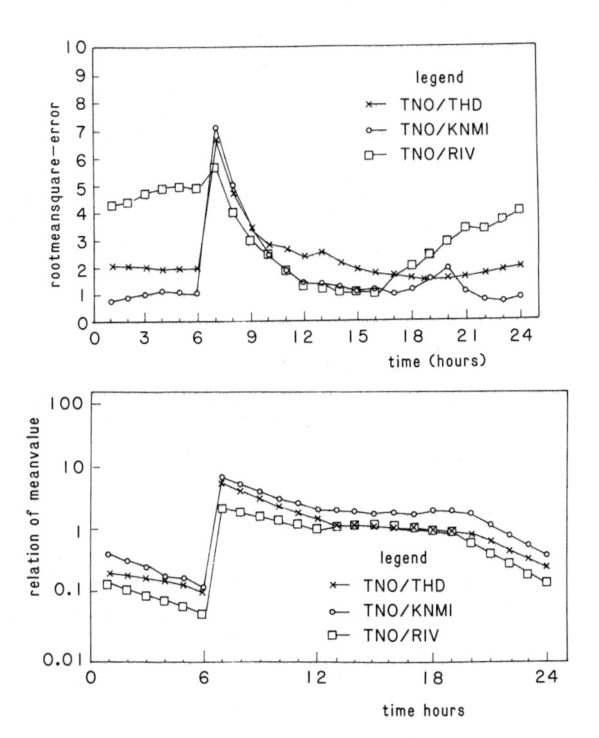

Fig. V.17 Statistical quantities TNO (SAI) / Models (phase II; episode β).

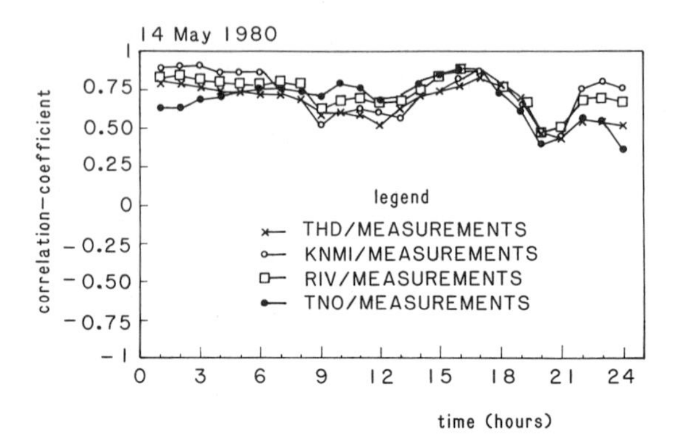

Fig. V.23 Correlation-coefficient of Models / Measurements (episode β; phase III).

Deposition amounts are the lowest for the KNMI output data. This is caused by the smearing out procedure mentioned in phase I (Figure V.16). The displaced position of the source is also shown clearly. The maximum surface concentration is shifted to the West.

Analysis of phase III:

The model output is characterised by two westward plumes originating from the Ruhr area. As observations in Belgium are lacking, this double structure cannot be confirmed. On the average, however, the model output correlates well with the available observations in the Netherlands (Figure V.23) during almost all the time of the day. Remarkable is the decrease in correlation for all models at sunset - 18.00 GMT. The temperature inversion which starts building up at sunset, stabilizes the boundary layer. This prevents the downward flux from the high stacks. The models do not describe this effect before 22.00 GMT (Figure V.21), while the measurements show it already from 18.00 GMT.

In Figure V. 18 - V.19 the abrupt fumigation of the THD and TNO (SAI) model is demonstrated. Apparently low vertical diffusion coefficients have been used during night time. In the morning hours the accumulated mass in the upper layer is mixed down very rapidly. The observations, KNMI and RIV data do not show this feature. There is also a time lag between the fumigation moment of the THD- and TNO (SAI)-model.

(text continued on page 249)

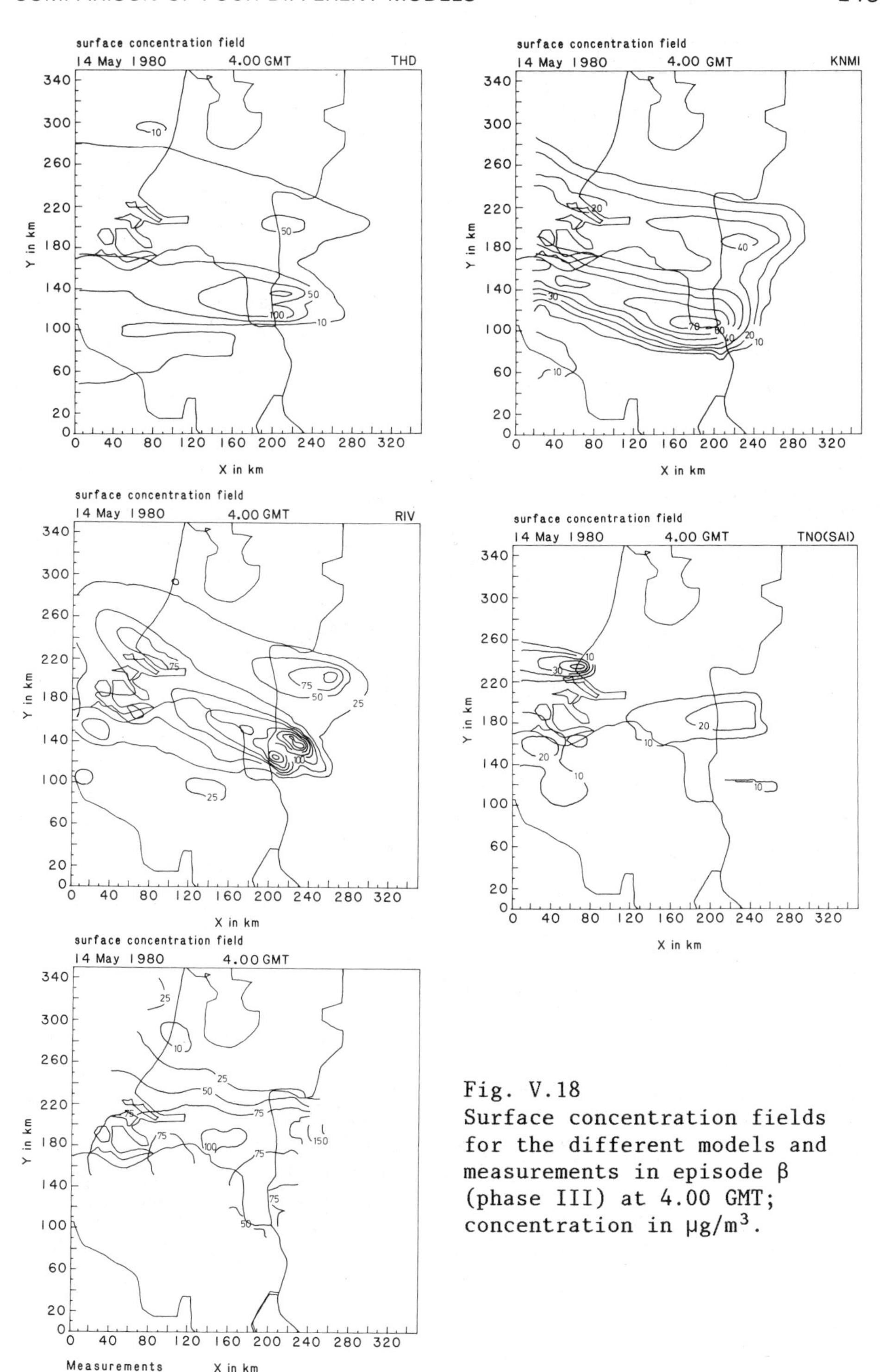

Fig. V.18
Surface concentration fields for the different models and measurements in episode β (phase III) at 4.00 GMT; concentration in μg/m³.

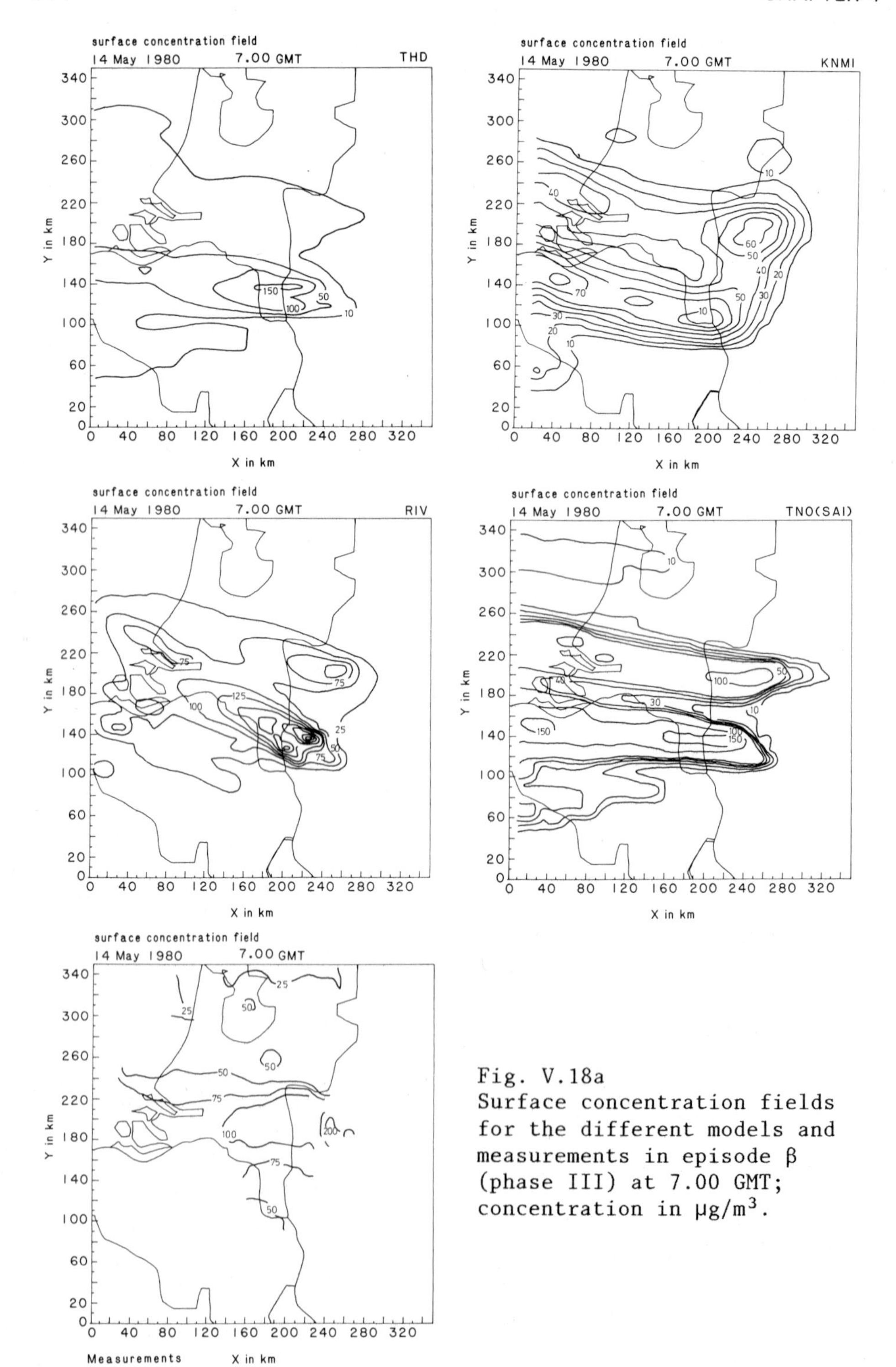

Fig. V.18a
Surface concentration fields for the different models and measurements in episode β (phase III) at 7.00 GMT; concentration in μg/m³.

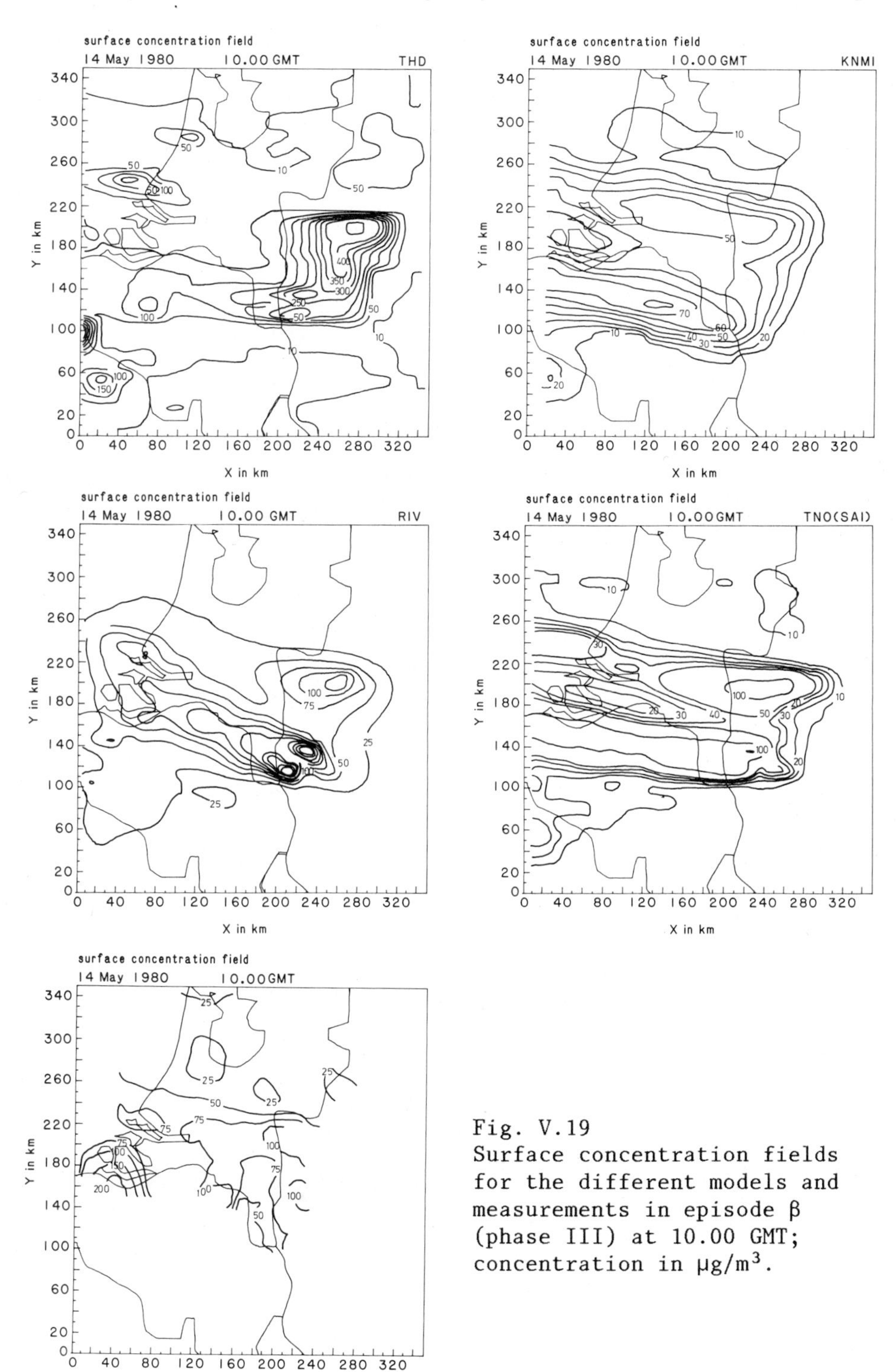

Fig. V.19
Surface concentration fields for the different models and measurements in episode β (phase III) at 10.00 GMT; concentration in $\mu g/m^3$.

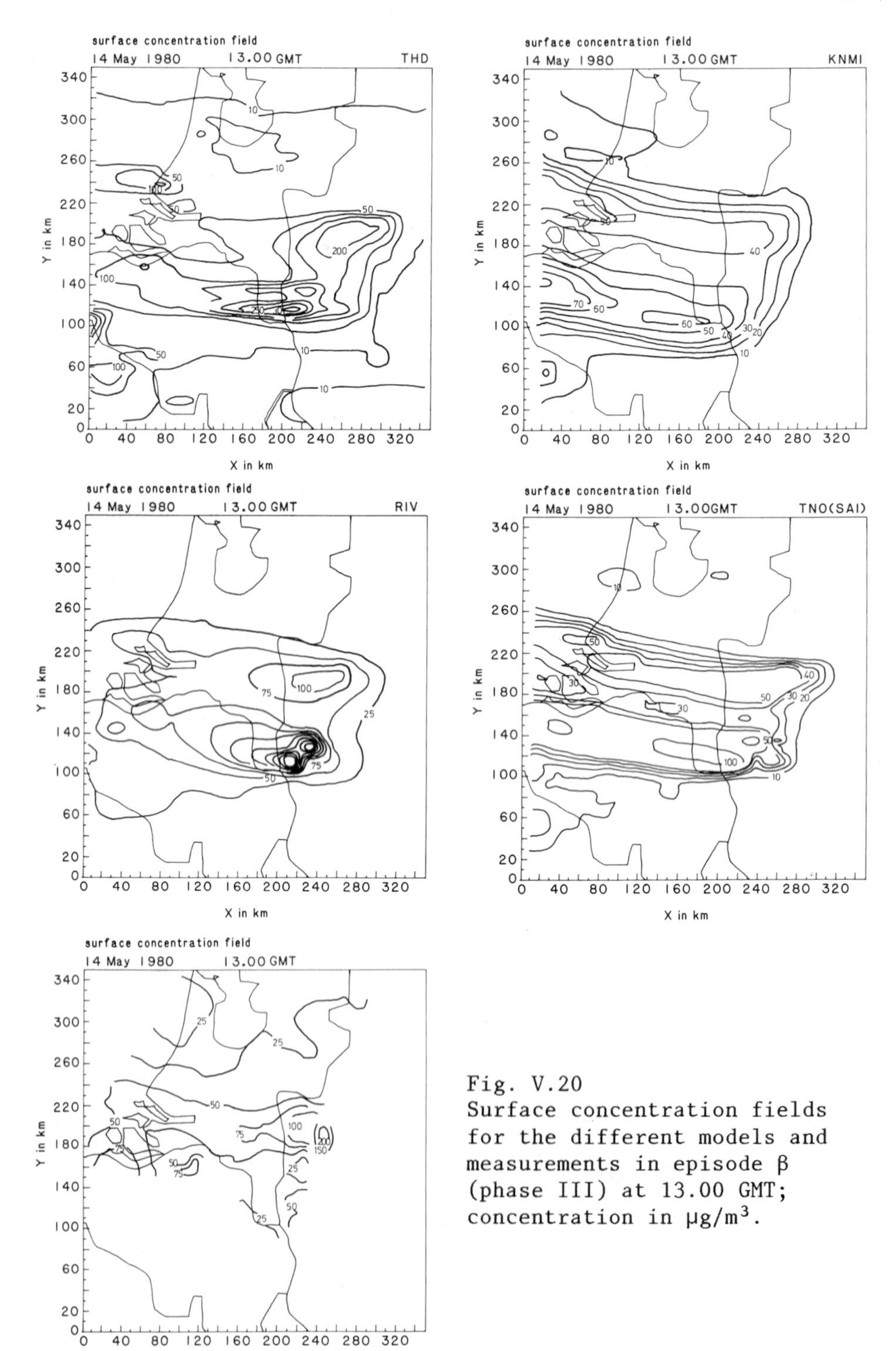

Fig. V.20
Surface concentration fields for the different models and measurements in episode β (phase III) at 13.00 GMT; concentration in μg/m³.

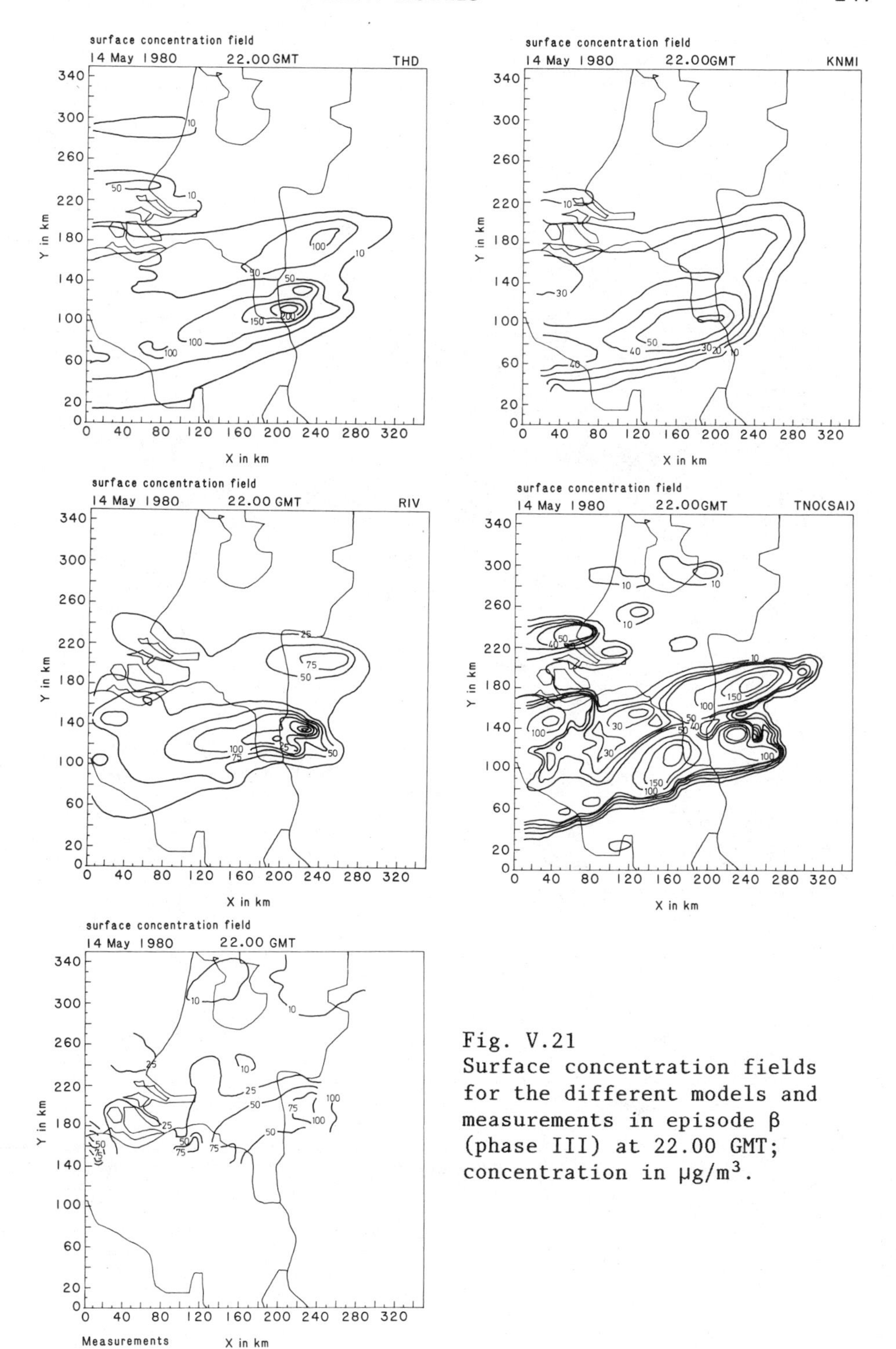

Fig. V.21
Surface concentration fields for the different models and measurements in episode β (phase III) at 22.00 GMT; concentration in μg/m³.

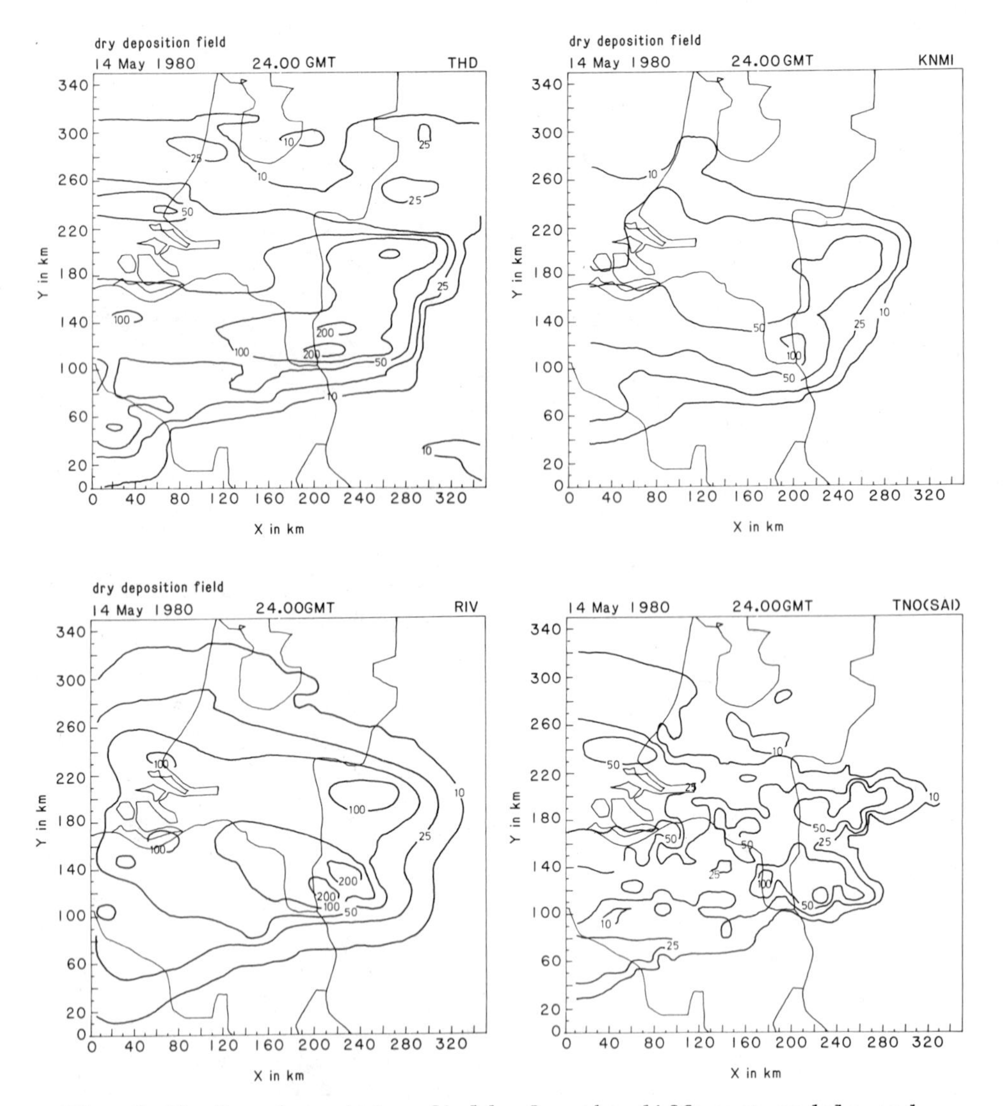

Fig. V.22 Dry deposition fields for the different models and measurements in episode β (phase III) at 24.00 GMT; deposition in mg/m^2.

The accumulated dry deposition fields reflect the fact that the meteorological situation is dominated by advection. The patterns are very much the same with a correlation coefficient about 0.8 (Figure V.22).

The output data of the THD- and RIV-model show twice as much deposited mass as the other two. This can be explained by the source handling (low values near the source) for the KNMI model and the relatively small deposition velocities used in the TNO (SAI) model.

V.c Analysis of episode γ (1981)*

Analysis of phase II

The depression which passes the central region of the Netherlands on December the 15th causes a strong easterly flow in the northern half of the region, during the night and the morning hours. Concentration patterns of all models extend in westerly or slightly south-westerly directions.

The concentrations predicted by the RIV-model are considerably lower until 10.00 GMT because in this model a nocturnal inversion (z = 150 m) inhibits downward diffusion of material from the source. This was apparently not the case in the THD and KNMI-models. This is also shown in the concentration fields of 7.00 GMT (Figures V.24). After 10.00 GMT the mixed layer height in the RIV model increases resulting in a fumigation of the surface layer. The concentration rises sharply exceeding the values calculated by the other two models. (Figure V.25).
Towards the end of the day wind velocity and mixing height increases again, resulting in generally lower concentrations, with an exception for the THD model (Figure V.26).
The mutual correlation in the concentration pattern rises also. The correlation between THD and RIV attains a value of 0.75 towards the end of the episode. The correlation of the KNMI model with the two other models, however, remains low (0.30-0.35). It should be noted that in the KNMI model the source position was slightly shifted with respect to the two other models. The correlation coefficient is particularly sensitive for this dislocation in single source comparisons (Figure V.30).

---

* For this episode TNO data were not available, due to time and funding constraints.

(text continued on page 254)

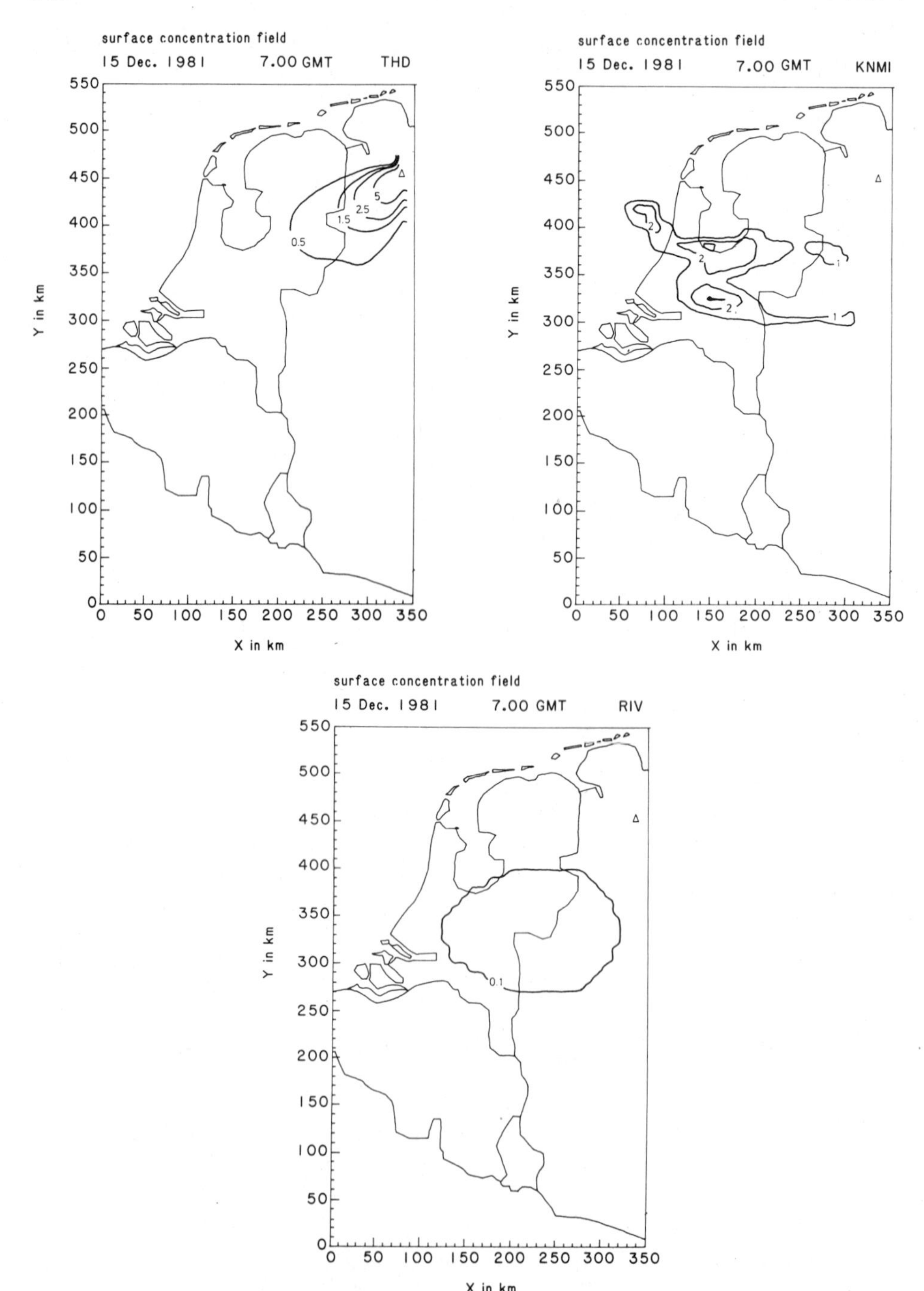

Fig. V.24 Surface concentration fields for the different models in episode γ (phase II) at 7.00 GMT; concentration in $\mu g/m^3$.

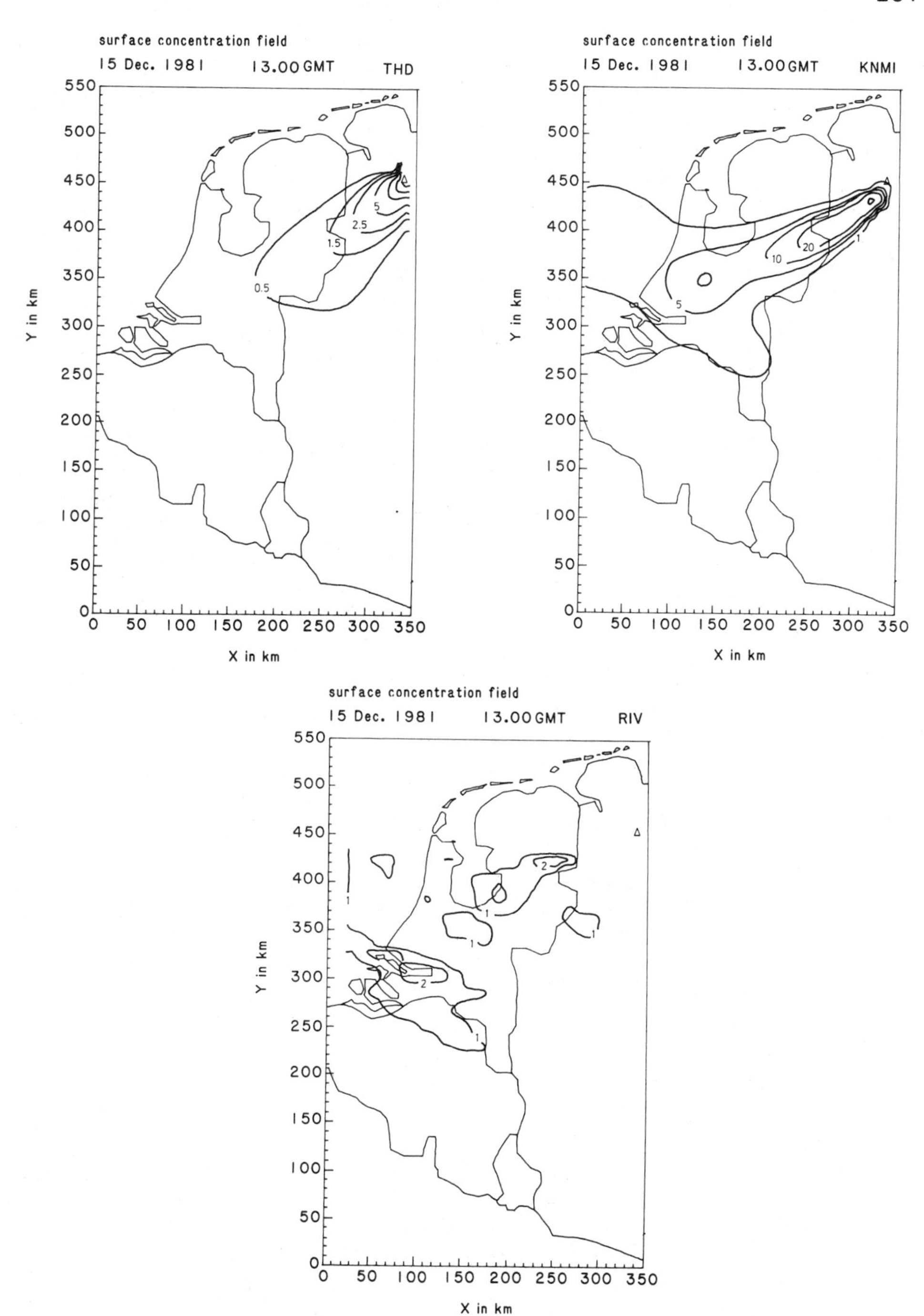

Fig. V.25 Surface concentration fields for the different models in episode γ (phase II) at 13.00 GMT; concentration in $\mu g/m^3$.

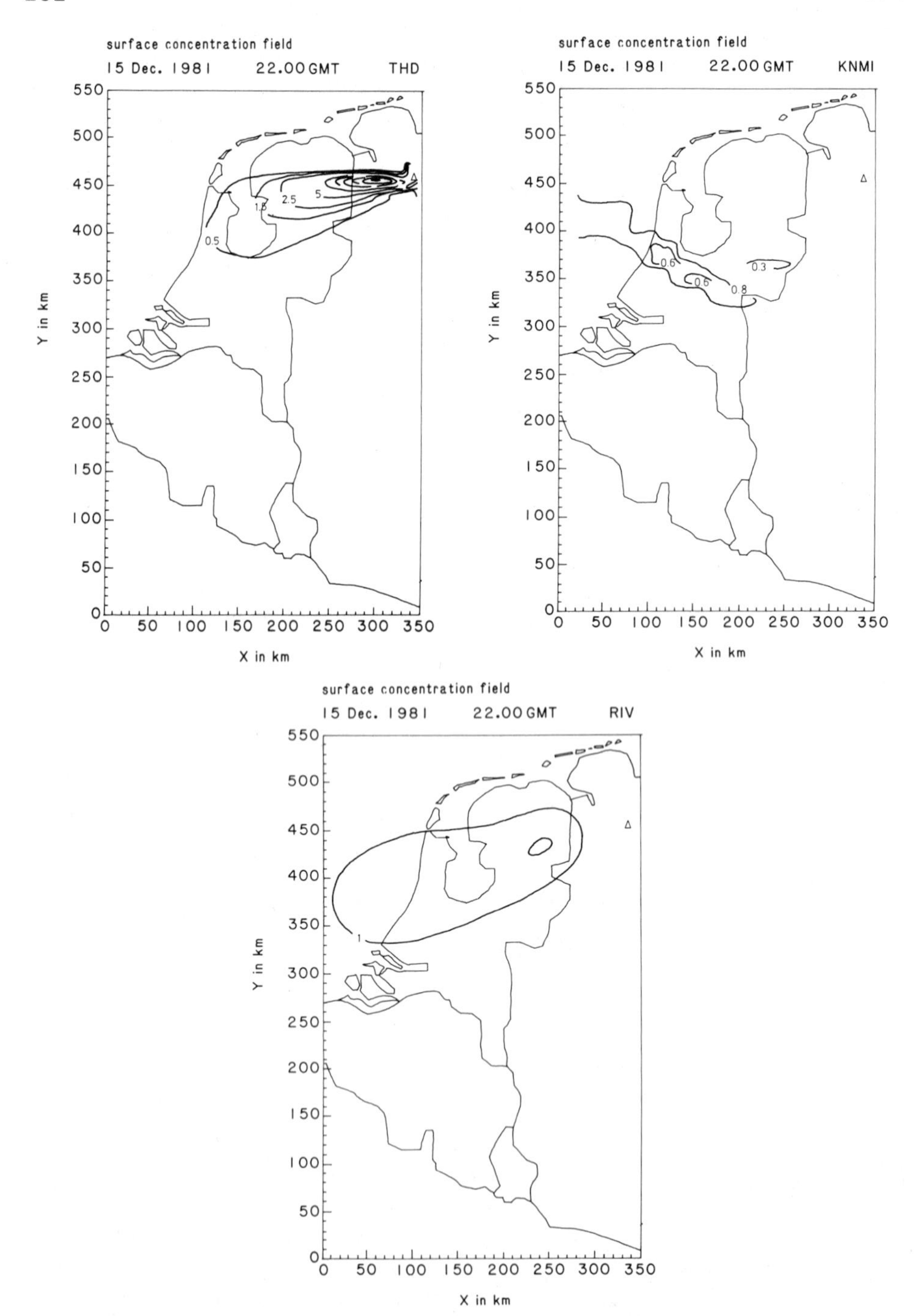

Fig. V.26 Surface concentration fields for the different models in episode γ (phase II) at 22.00 GMT; concentration in μg/m³.

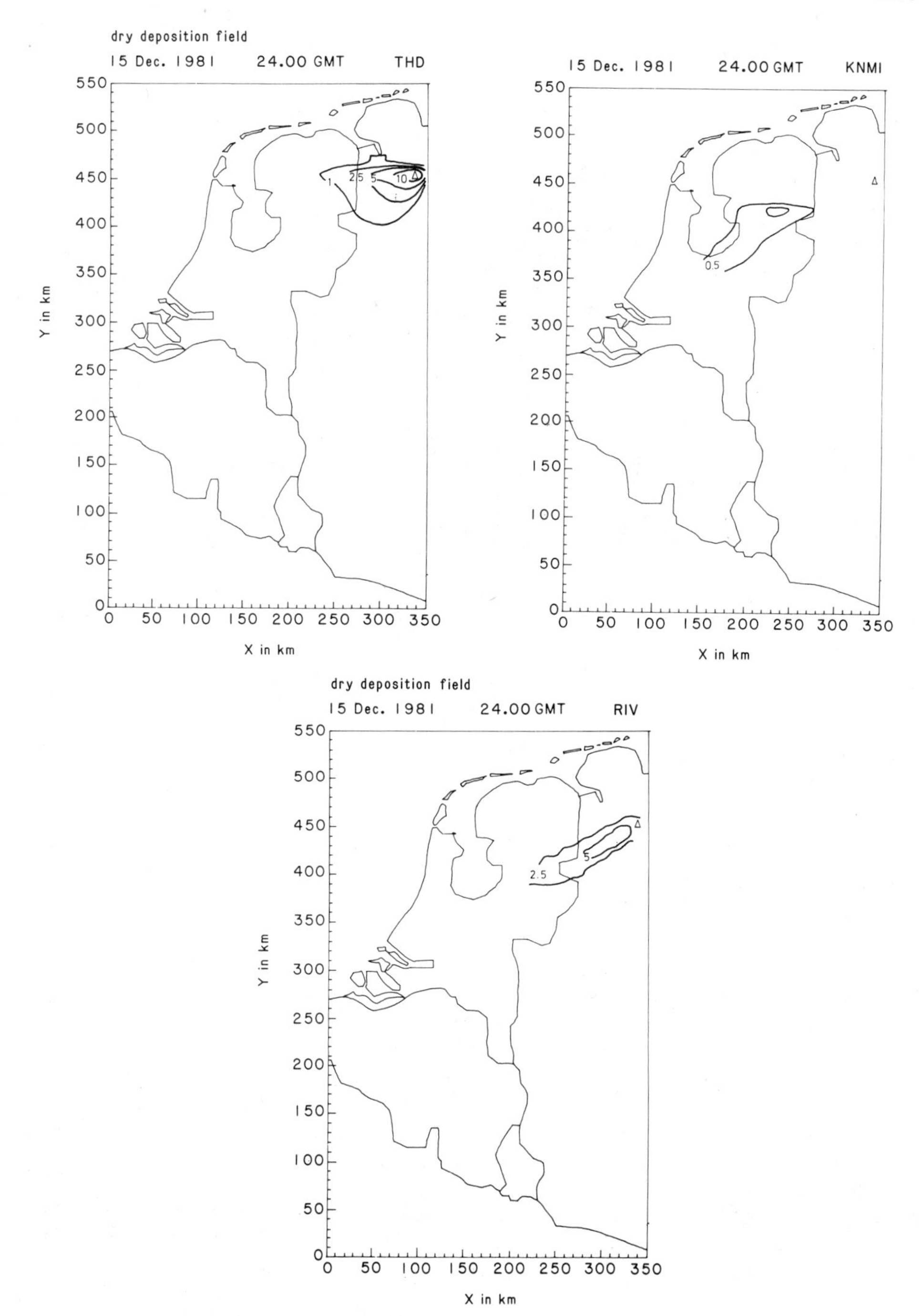

Fig. V.27 Dry deposition fields for the different models in episode γ (phase II) at 22.00 GMT; deposition in mg/m^2.

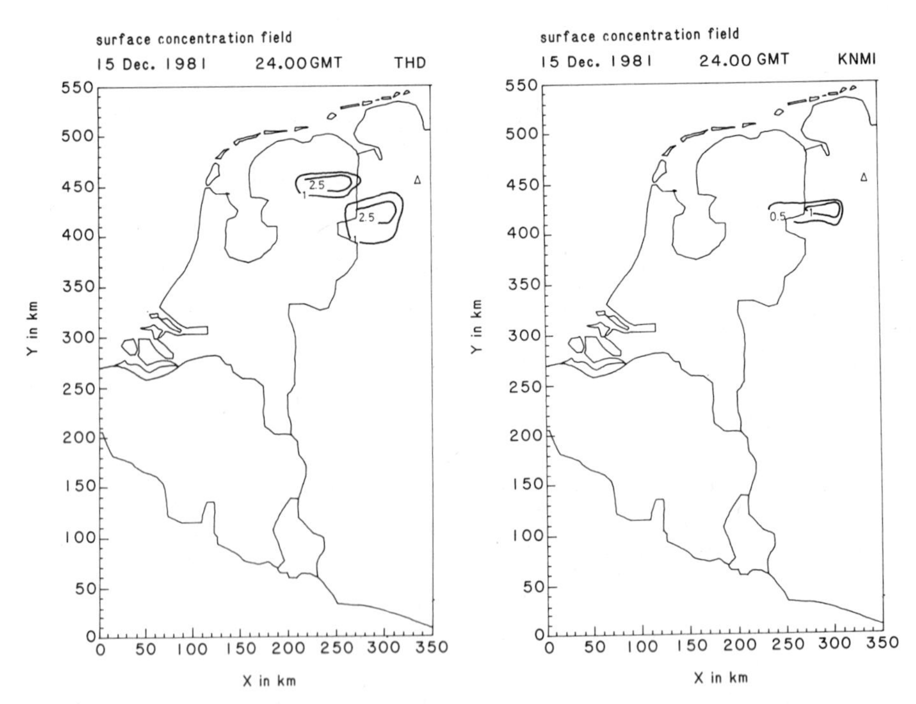

Fig. V.28 Wet deposition fields for the different models in episode γ (phase II) at 24.00 GMT; deposition mg/m^2.

In this episode the model results are often different from each other. This may be partly ascribed to the very complicated meteorological situation during this episode. There are, however, also common features. All models show initially an advection pattern, which moves gradually to the south-west. Soon after that the concentration field is trapped by the rotating wind field associated with the depression moving over the area. This wind field distorts and bifurcates the concentration field into two distinct sections, one moving towards the west and the other extending eastwards. This is most clearly shown by the KNMI model (Figure V.24) and somewhat less pronounced by the RIV model (Figure V.25). It is not apparent in the THD model concentration field.

The average concentration values of the KNMI model are towards the end of the period much lower than those of the other models. This is mainly caused by the large mixing height in the KNMI model.

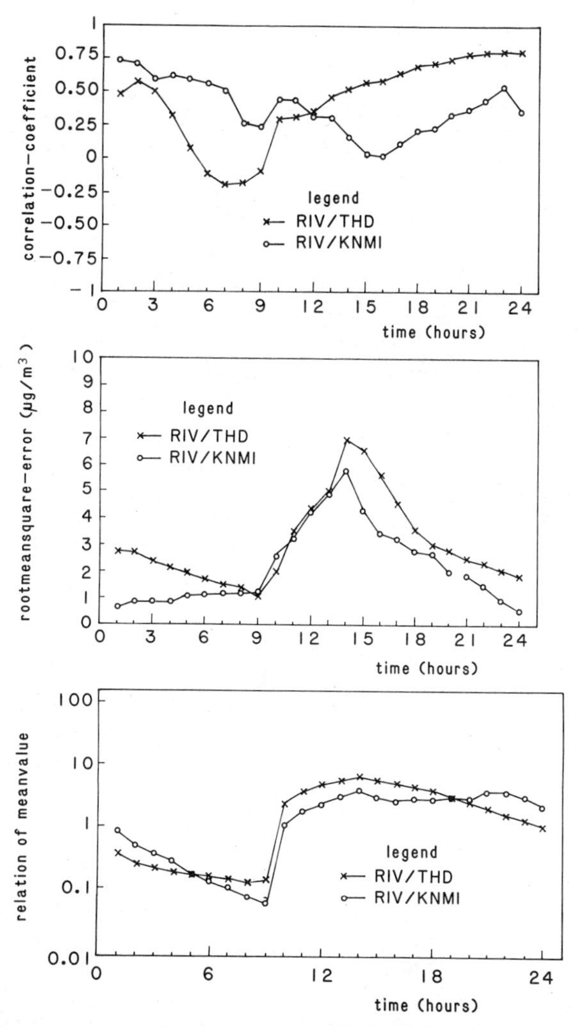

Fig. V.29 Statistical quantities RIV/models. (episode γ; phase II)

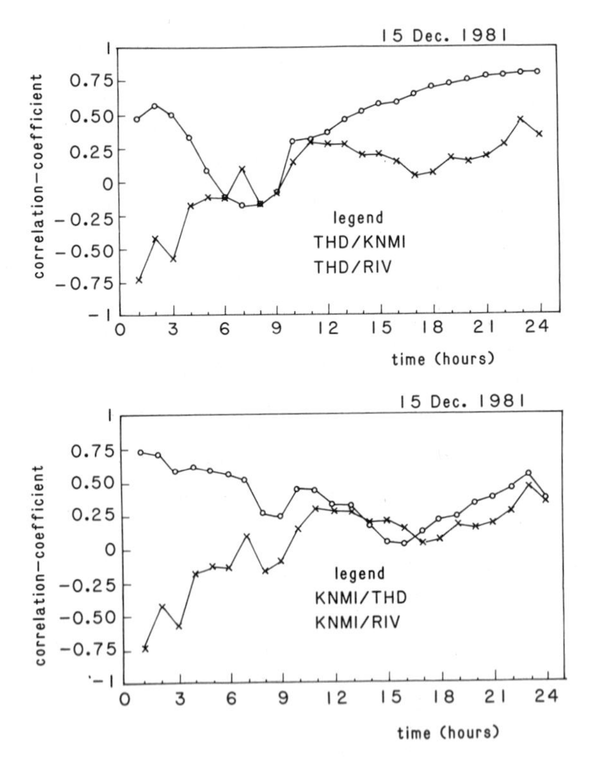

Fig. V.30 Correlation-coefficients of THD/models, resp. KNMI/models (episode γ; phase II).

The dry deposition is largest in all models in a region south-west of the source. The THD field is limited to a relative small area near the source (Figure V.27). When the shifted source location, towards the west in the KNMI calculations, is accounted for all patterns are similar. There are, however, differences in magnitude mainly due to the differences in concentration fields of the various models. The KNMI produces the lowest deposition values (ca. 0.5 mgm^{-2}). The RIV and THD models produce values of ca. 2 mgm^{-2}. (Deposition close to the source (< 20 km) is neglected).

Wet deposition is only calculated by THD and KNMI. Also here the KNMI model produces the lowest values (ca. 0.5-1 mgm^{-2}), whereas the THD model values range from 1-2.5 mgm^{-2}. In the THD model wet deposition occurs in an appreciably larger area south west of the source (Figure V.28).

Analysis of Phase III

During the first hours of the episode all models give high concentrations in the southern half of the region (Fig. V.31). KNMI values are the lowest (25 μgm^{-3} or less). RIV and THD show values ranging from 50-100 μgm^{-3}. In this region the observations range from 10-50 μgm^{-3}. (Figure V.32). In the northern half of the region high concentrations are observed with values over 150 μgm^{-3}. These are not predicted by any of the models. There are strong indications that this is caused by intrusion of heavily polluted air over the boundary of the model region (Figure V.32). Incidental releases of SO_2 from the NW Germany natural gasfields are often observed at low ambient temperatures when the natural gasproduction exceeds the capacity of the desulphurization installations. These emissions are not included in the inventory. This explains why the correlations between measurements and models are initially not significant.

Towards noon the models produce chaotic concentrations patterns with maximum values over the Ruhr area, Belgium and the south western part of the Netherlands. This is in fair agreement with the measured pattern (Figure V.34). Correlation coefficients between models and measurements are fair showing correlations between 0.7 and 0.85 after 13.00 GMT (Figure V.33). The correlation of the KNMI model with measurements is slightly less, which is presumably caused by the much smaller subset of correlation pairs.

The intercorrelation between models is generally fair (higher than 0.6). Also during the afternoon and night concentrations are highest in the southern half of the area, both in models and in the observations (Figure V.35). The dry deposition is concentrated as expected in the areas with the highest mass concentrations. There is an excellent agreement between the three models giving maximum values in the Ruhr area, Belgium and the Rijnmond area of respectively $\cong$ 100, 50-100 and 25-50 mgm^{-2} (Figure V.36). The mutual correlations of the concentration patterns vary between 0.80 and 0.85.

The wet deposition fields are only available for the THD and KNMI model. The THD model shows a very patchy pattern with values exceeding 100 mg m^{-2} over large parts of Belgium and Western Germany (Figure V.37). The KNMI model also depicts maximum values in these areas but they do not exceed 25 to 50 mg m^{-2}, the difference between these two models being approximately a factor of 3. This issue clearly is not yet fully resolved and should be looked into further.

VI. SUMMARY

In the beginning of the comparison it was decided that no flux across the boundaries and no initial background concentrations

(text continued on page 260)

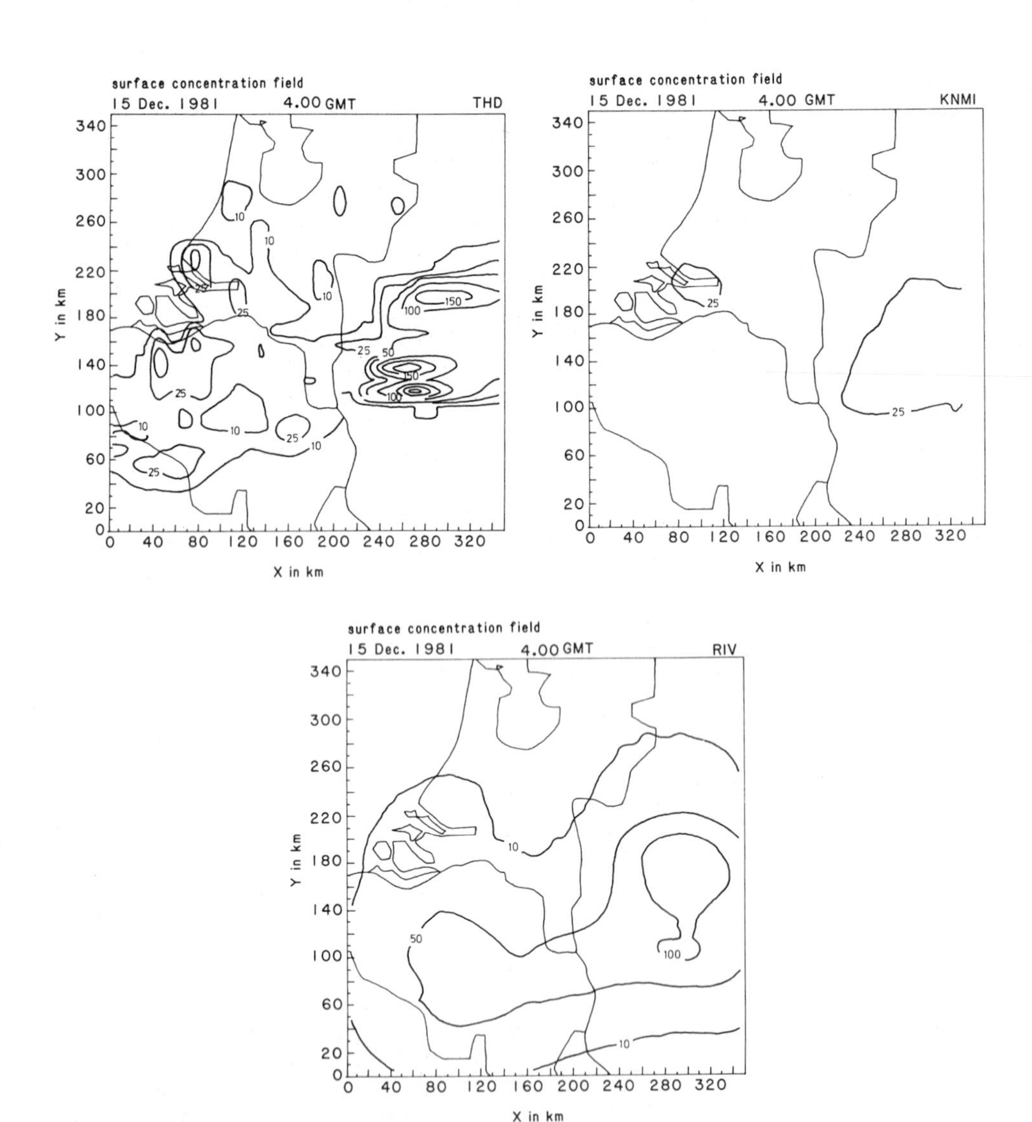

Fig. V.31 Surface concentration field for the different models in episode γ (phase III) at 4.00 GMT, concentration in $\mu g/m^3$.

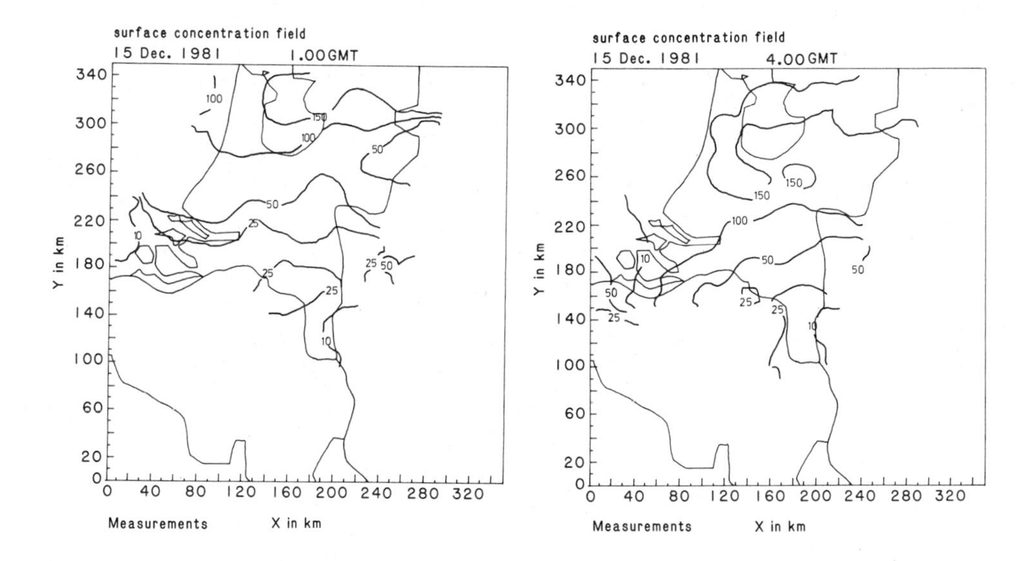

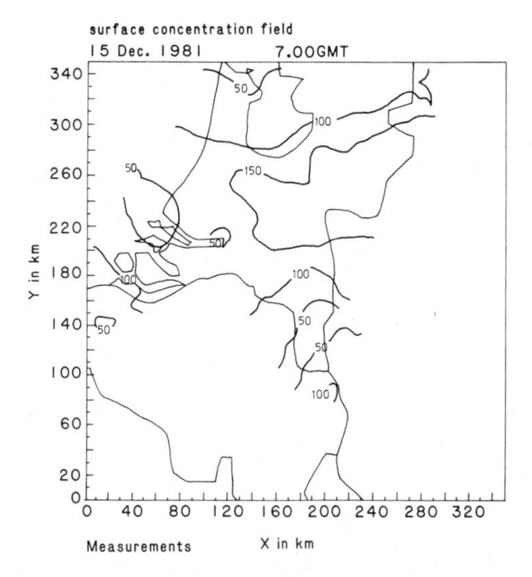

Fig. V.32 Surface concentration fields from measurements during episode γ (phase III), concentration in $\mu g/m^3$.

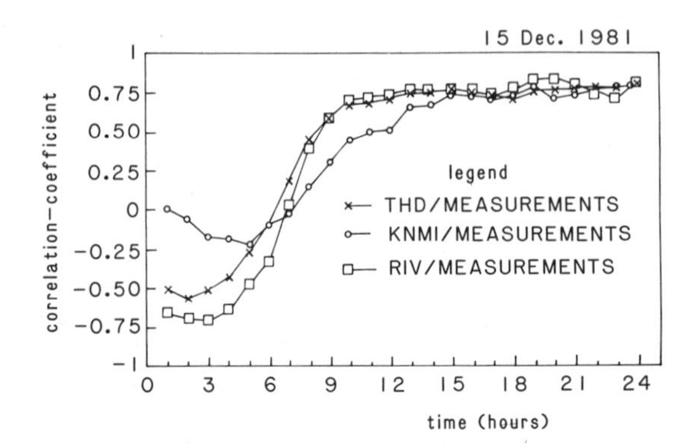

Fig. V.33 Correlation-coefficients of models/measurements (Phase III, episode γ).

should be considered in the calculations. This was decided because of the lack of suitable information on these quantities at that moment. It is repeated here that the comparison of model results with observed data does not therefore constitute a criterion for judging model performance. The intercomparison of the four different interregional models revealed the following facts which have been described in more detail in the former chapters and which are summarized here:

1. Four interregional models were compared as to their results. Three of them were of the Eulerian type, one was Lagrangian. They differed in many respects, and their characteristics are shown in Table II.1.

2. The intercomparison study was carried out in three phases. In phase I the numerical model procedures were evaluated. In phases II (one hypothetical point source) and III (with real emission inventory) model performance was compared under three different meteorological situations. The data have been discussed in detail in chapter IV.

3. In phase I many properties of the models were tested though not all. The different treatment of the horizontal diffusion in the models for example, was not explicitly tested since the horizontal extension of the plume was approximately of the order of one grid cell. But in general the results of phase I document clearly that all four models simulate the analytical solution(s) of case A and B to a sufficient degree of accuracy.

(text continued on page 264)

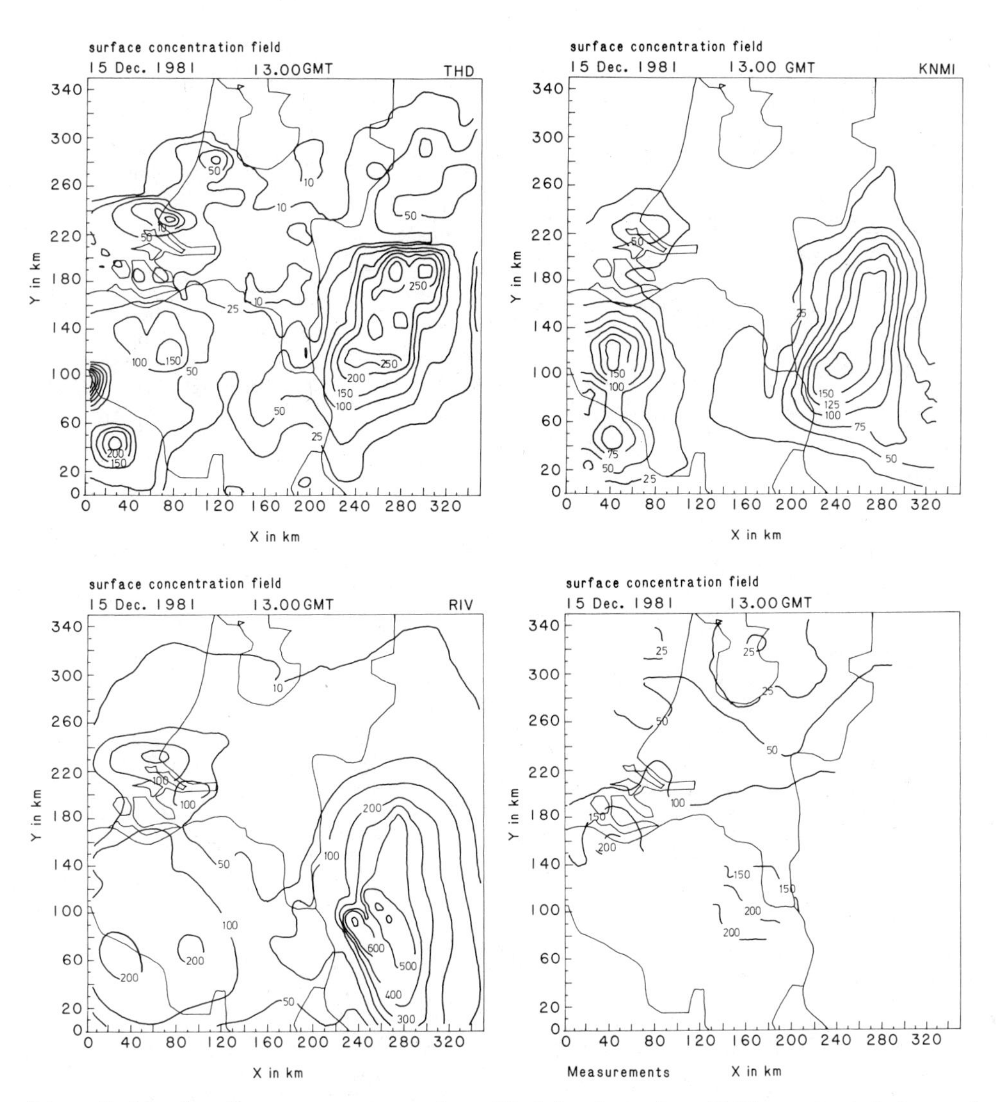

Fig. V.34 Surface concentration fields for the different models and measurements in episode γ (phase III) at 13.00 GMT; concentration in μg/m³.

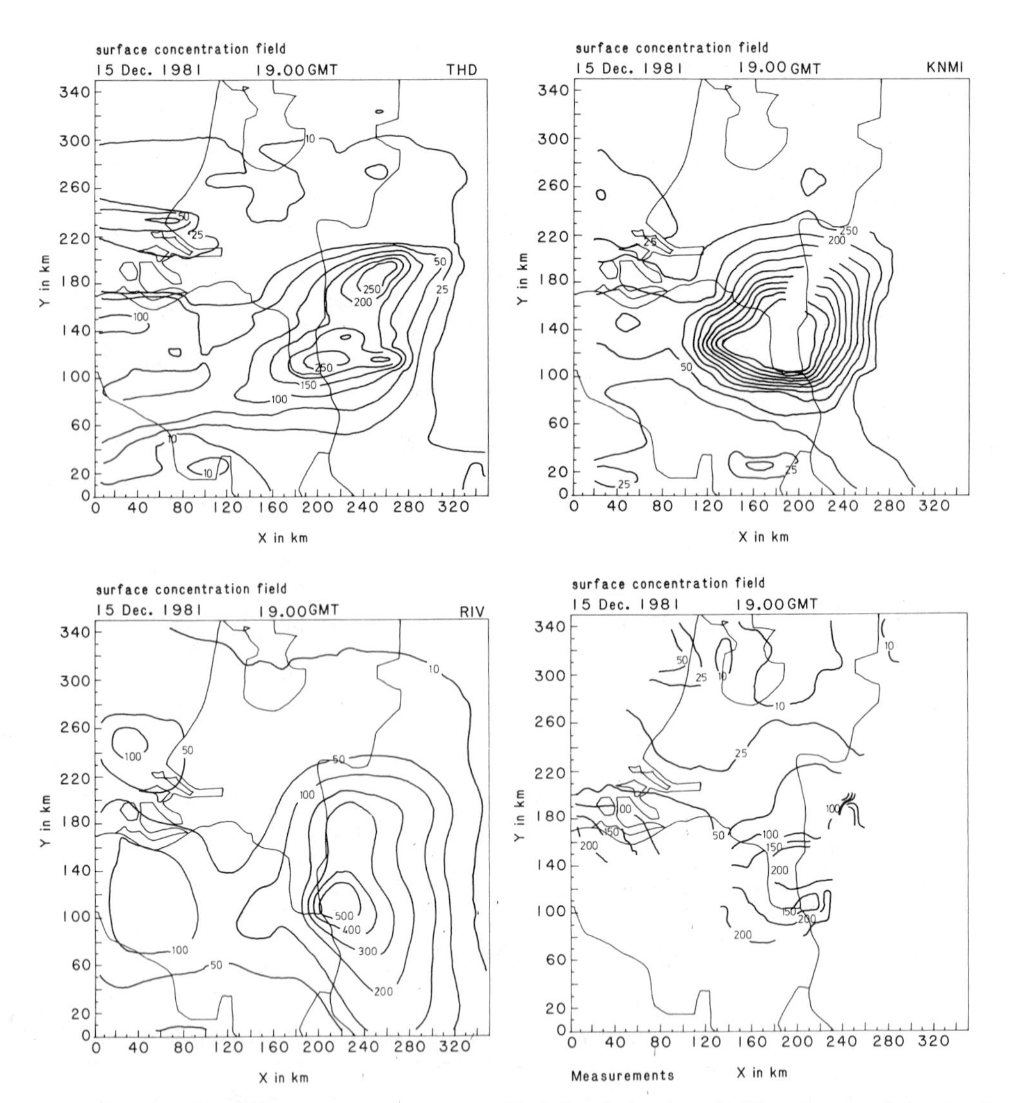

Fig. V.35 Surface concentration fields for the different models and measurements in episode γ (phase III) at 19.00 GMT; concentration in μg/m³.

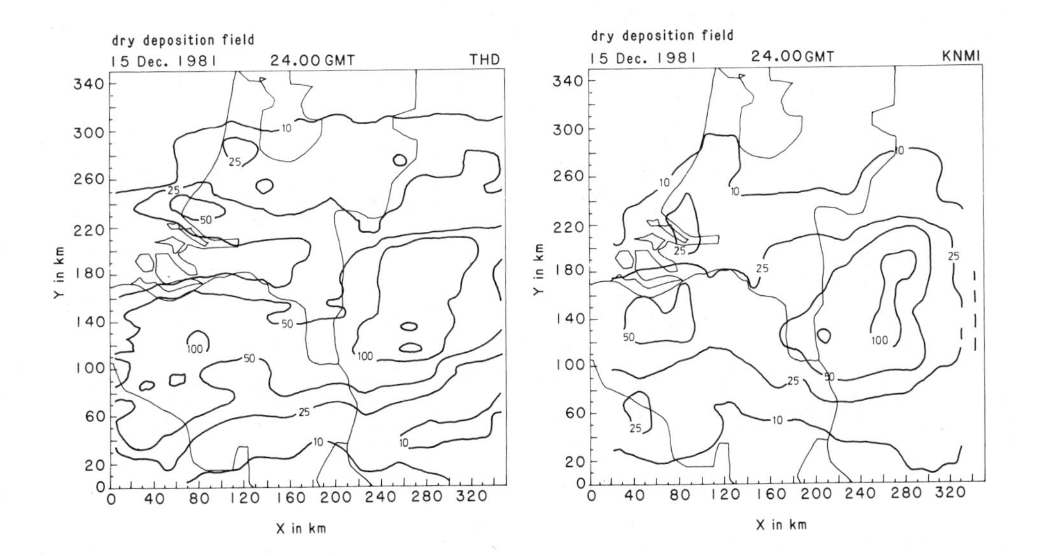

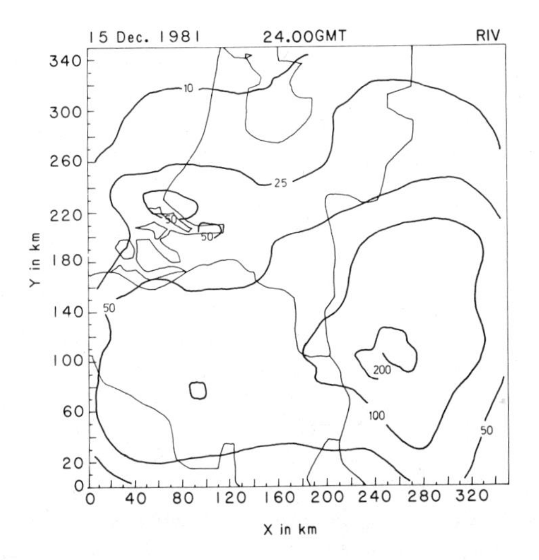

Fig. V.36 Dry deposition fields for the different models in episode (phase III) at 24.00 GMT; deposition in mg/m^2.

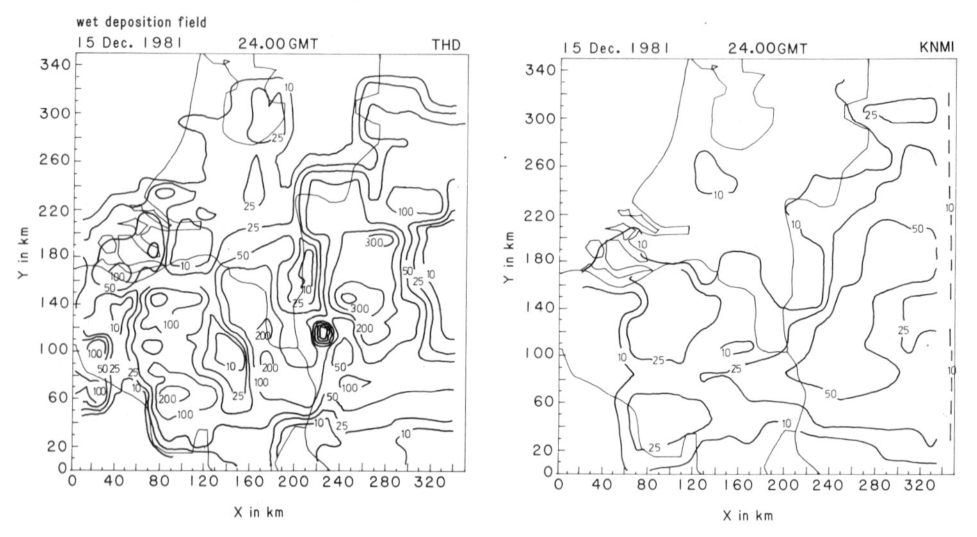

Fig. V.37 Wet deposition fields for two models in episode γ (phase III) at 24.00 GMT, deposition in mg/m².

The results for test case C (simulation of a non-stationary meteorological situation) show the differences between the models concerning the treatment of the vertical diffusion which are especially obvious in the beginning of the simulation. The correlation between the model results improves with time.

4. The differences between the models shown in phases II and III in the intercomparison study are not due therefore to the different model structures as such, but most likely due to the - sometimes completely - different processing of the input data (which were the same for every model) to the model core. This concerns especially the windfield, the vertical and horizontal diffusion, the mixing height and also the precipitation. It should be noted that RIV has used acoustic sounder data to estimate mixing heights, which data were not available for the other participants.

5. The results for phases II and III which have been described in detail in chapter V reveal the following features: (in the comparison the results were always compared for the second day of an episode)

a) Episode α (1979)

This is a stagnant period with parts of the area snow-covered. The analysis of episode α shows that the dispersion of pollutants in the beginning is dominated by vertical and horizontal diffusion rather than advection.

The results of the computations of phase II display the differences in the simulation of the vertical mixing (e.g. fumigation), the advection and the horizontal dispersion by the four models. Differences in vertical stability and mixing-height profiles result in large discrepancies from 0.00 hours to 10.00 hours, as illustrated by the concentration fields at 7.00 hours (Figure V.1). Horizontal dispersion and advection treatment at 22.00 hours finally leads to different geographical positions of the core of the plume (Figure V.4). The strongest horizontal dispersion is simulated by the RIV-model resulting in low gradients of the concentration field at 13.00 hours (Figure V.3) and 22.00 hours (Figure V.4). The results of the TNO(SAI)-model are characterized by box-shape plume patterns, having low gradients in the plume centre and strong gradients on the plume edges.
The KNMI- and TNO(SAI)-results show a decoupling of the ground concentration maxima (plumes) from the source as illustrated by the fields of 7.00 and 22.00 hours (Figures V.1 and V.4).

The very detailed deposition patterns of KNMI and especially TNO(SAI) reflect the more subtile treatment of dry deposition as a function of land use. Highest deposition rates are computed by RIV, the other models compute about the same values. This is explained by the presence of the snow cover, which was not accounted for in this version of the RIV-model.

The results of phase III show that in relation to the weak advection during the first half of the day and the more stationary west-southwest transport during the second half of the day the correlations between models and measurements improve towards the end of the day (Figure V.13). Figure V.7 shows that a puff was observed with concentrations between 500 and 1000 μg/m^3 SO_2 in the Rotterdam-Rhinemouth area. This puff is firstly transported in the east-southeast direction and then after 13.00 hours advected to the east-northeast direction. None of the models simulates this behaviour adequately, which explains the low correlations with the measurements in the beginning of the day (Figure V.13). During the second half of the day when the pollution is advected to the northeast, the correlations between models and measurements improve.

b) Episode β (1980)

This episode is dominated by a strong easterly wind. The mixing height is always greater than 900 m. The results of phase II show that the models produce similar patterns (Figures V.14, V.15). The largest difference occurs between the TNO(SAI)-results and the ones of the other models during the night. This seems to be due to the fact that the procedure used by TNO(SAI) to estimate mixing heights leads to the prediction of concentration levels that are significantly lower than those actually measured. After 6.00 GMT the TNO(SAI)-results show the same pattern as the other models.

For phase III all models produce two westward plumes originating from the Ruhr area. As observations in Belgium are lacking, this double structure cannot be confirmed. In general, however, the model results correlate surprisingly well with the available observations in the Netherlands (Figure V.23) during the day. But at 18.00 GMT the models seem to have difficulties to simulate the stabilization of the surface and boundary layer. They describe this effect not before 22.00 TMT (Figure V.21), whereas the measurements show it already at 18.00 GMT.

The THD- and TNO(SAI)-results show an abrupt fumigation between 4.00 GMT and 10.00 GMT illustrated by Figures V.18 and V.19. This pattern is probably due to the low vertical diffusion coefficients used at night. The THD-model produces very high values (~ 400 $\mu g/m^3$).

The patterns of the dry deposition fields are very much the same for all models. The results of the THD- and RIV-model show about twice as much deposited mass as the other two. This can be explained by the low values near the source for the KNMI-model and the relatively small deposition velocities used in the TNO(SAI)-model.

c) Episode γ (1981)

For this episode only the results of RIV, THD and KNMI were available.

In phase II the calculated concentration pattern of the models extend in westerly or slightly south-westerly direction.
Until 10.00 GMT the RIV-results are lower than the other ones. Afterwards the increase of the mixed layer height results in a concentration rise (Figure V.25). Towards the end of the day wind velocity and mixing height increase again, this time resulting in generally lower concentrations (lowest for KNMI-model because of the use of a large mixing height), with an exception for the THD-model.

Besides the many differences between the model results there are also common features. All model results initially show an advection pattern, which moves gradually to the south-west. Soon after the concentration field is trapped by the rotating wind field associated with the depression moving over the area. This wind field distorts and bifurcates the concentration field into two distinct sections, one moving towards the west and the other extending eastwards. This is very clearly shown by the KNMI-results (Figure V.24) and somewhat less pronounced by the RIV-results (Figure V.25). It is not apparent in the THD-concentration field. The pattern of the deposition fields are very similar. But there are differences in the magnitude of the values. The KNMI-model produces the lowest deposition values (ca. 0.5 mg m^{-2}) and the other two models have values of ca. 2 mg m^{-2}.

Wet deposition is only calculated by THD and KNMI. Here the KNMI-values are also smaller (ca. 0.5 - 1 mg m^{-2}) than the ones from THD (1 - 2.5 mg m^{-2}).

The results of phase III and the measurements seem to indicate that in the beginning of the period an influx of pollutants occurs from an area which was not covered by the emission inventory. Since all models due to lack of data did not include fluxes across the boundaries of the computation area, the agreement between model results and observations could not be good during this time period (see Figure V.32). Later on in the day also a fair agreement between models and observations was achieved for this complicated case (Figure V.33 and V.34). There is an excellent agreement between the dry deposition fields of the three models in giving maximum values in the Ruhr area, Belgium and the Rhinemouth area of respectively $\simeq$ 100, 50 - 100 and 25 - 50 mg m^{-2} (Figure V.36).

The wet deposition fields of the THD- and KNMI-models show maximum values over large parts of Belgium and Western Germany. But the difference in magnitude between the models is approximately a factor of 3.

d) Final remarks

There are certain shortcomings of the model applications in this study which can be described as follows:

- The KNMI-model displaces the sources in phases II and III away from the boundary into the computation area. Furthermore, a point source is simulated by distributing this source over a number of grid points.
- The procedure used to estimate mixing heights input to the TNO(SAI) Model yielded relatively low nocturnal values causing emissions from elevated sources to be trapped aloft; observed concentrations suggest a greater degree of mixing to the ground.

- The THD-model produces a too strong fumigation effect during the 1980-episode.
- The RIV-model does not simulate the fumigation effect which the other models show during the 1979-episode.

VII. CONCLUSIONS AND RECOMMENDATIONS

The outcome of this study which compares results of four completely different interregional episode models working with the same input data set has led us to the following conclusions:

A. The differences in the output data of the four models are mainly due to the differences in processing the raw meteorological data, i.e. the way of transforming the routine meteorological information into the input needed by the model core.

B. None of the four models shows a clear superiority over the others: one treats a certain meteorological situation better than the others, but is less good at an other occasion. From this we conclude that with the meteorological and emission input data used for this study the overall performance of the investigated models is practically the same, regardless of the degree of their complexity and refinement.

C. Furthermore, for model calculations on this scale we conclude that the (three-dimensional) initial conditions and the flux of pollutants across the boundaries (also as a function of height!) must be taken into account.

D. From all of the meteorological input parameters the mixing height and the wind field have the largest influence on the model results. Here certainly more research is required to find methods to infer reliable estimates of these quantities from routine meteorological data.

E. The accumulated dry deposition fields reflect the concentration fields to a large extent and show the differences (Figure V.11) between the models in the treatment of the dry deposition concerning land use and deposition velocities. The comparison of the dry deposition values with measurements is not possible because of lack of data.

F. A comparison between the wet deposition for the one period has been limited to only two models and the results - as is the case for dry deposition - cannot be compared with the measurements. There is, however, a large difference between the two model results which seems to be due to the different treatment of the input data (e.g. different interpolation of the precipitation observations) and the use of different washout- and rainout coefficients.

G. Since no effort has been made to model realistic concentrations (no background concentration, no initial concentrations and no flux across the boundaries have been included) and nevertheless the correlation coefficients when comparing with ambient data are reasonably high during most of the time, there are reasons to expect that model performance can be substantially improved when these effects are taken into account.

H. It is recommended that the present data set of emission and meteorological data is used for other model comparison, since a number of models have already been using these data in the study presented, and some experience with this data set is already available.

APPENDIX

At Friedrichshafen an expert workshop was held at Dornier on December 5 and 6, 1983 to discuss the first draft of the report of this intercomparison study. Seven experts from five countries were present, Dr. A. Eliassen - Norwegian Meteorological Institute; Dr. B. Fisher - CERL U.K.; Dr. F. Nieuwstadt - KNMI the Netherlands; Dr. S. Reynolds - SAI USA; Dr. B. Smith - Meteorological Office U.K.; Dr. R. Stern - FU Berlin; Dr. R. Yamartino ERT USA. The detailed comments made during this workshop were used to improve the report. The most important general comments written down by the experts are summarized in this appendix.

According to phase 1 the following comments were made.

- A particularly useful part of the study is the comparison of the models for some simple cases for which an analytical solution to the diffusion equation exists. In this way the model codes have been tested objectively. A minor point of criticism is that the test solutions were all done for point sources. None of the models, except possibly the RIV model, is well suited to simulate the dispersion of a point source. This is in particular the case for the KNMI-model which behaves poorly in the first stage of point source dispersion because it distributes the source over neighbouring grid squares. However, these regional models are not meant to model point sources accurately. In general, they require a separate point source module. Therefore, a point source simulation must be considered as a rather severe test.

Another expert commented on the results of phase 1 as follows:

- I would have liked more emphasis to have been laid on phase 1. It seems essential with large numerical packages to provide exhaustive tests of the numerical procedures. The examples chosen gave encouragement to the modellers to proceed with phase 2, but phase 1 work had already shown possible errors. In particular the influence of artificial diffusion was not quantified specifically and might be expected to cause severe problems with the lateral dispersion of material from a point source. The modelled dispersion may occur too rapidly near to the source. Problems with mass conservation had also become apparent. An additional test, for example, would be to apply the numerical schemes to an idealized flow field representing a simple rotation.

  Considerable savings are to be obtained in optimizing the numerical methods. Of course the time scale of this study would not have allowed detailed treatments. Furthermore I would strongly recommend that further work should be carried out in this area, from which the potential benefits could be very great.

In connection with phases 2 and 3 the following comments were made.

- I am in full agreement with the Conclusions of the report in connection with phases 2 and 3. In phase 2 one might have compared directly the derived wind fields used in the models (an other area for further work). No single measure for testing the agreement between models emerged as clearly superior in the discussion. However, I am not sure if such a measure is necessary in practice. Phase 3 was involved in the simulation of three episodes, two of which in meteorological situations which are notoriously difficult to treat. In further work case studies of simpler meteorological situations should be considered first and the difficult situations only dealt with later on.

Additionally one expert mentioned a basic problem very explicitly.

- I think all of us present at the Meeting clearly saw how particularly important the vertical profile of stability was to the results, in particular the ground level concentration field and the resulting dry deposition field. Where, on any occasion, this profile seemed to be well represented in one or more models, the comparison between these models' results and the observed concentration fields seemed very good indeed (excluding those situations which were strongly influenced by either the initial state or by transboundary fluxes). The models seemed to differ from each other more radically in this one respect - how they determined the mixing height and how they represented stable (mainly nocturnal) conditions - than in any other. Different mixes or radiosonde data, acoustic sounder data and theoretical - empirical formulae were involved. I think there is a great deal to be learnt here, especially as to how to treat the night-time situation. My first reaction was to applaud the way the mixing depth was treated by the KNMI-model, which seems to reflect our best theoretical understanding (even if I believe they tend to overestimate the depth of the neutral boundary layer by using 0.3 u_*/f rather that 0.2 u_*/f - a belief which is supported by KNMI's consistent underestimation of ground level concentrations). Nevertheless it seems eminently sensible to use acoustic sounder data like RIV has done and radiosonde data (although in the latter one obviously has to beware of missing the true top of the boundary layer which is often marked by a very small temperature jump, which may be missed by comparison with larger jumps at synoptically induced inversion at greater heights). The real problem is to combine the observed data and the theoretical predictions in the most sensible way, and I would guess that some sort of "choose the minimum value of the set" approach might be the most promising. In summary, both the daytime and nighttime stability profiles and the way these affect diffusion should be considered very carefully, and maybe in this respect some of the rather definitive field studies involving tracer releases, and detailed boundary layer measurements recently carried out in Idaho should be very valuable.

- The wet deposition fields are an other area where inter-model variability is bound to be large. On a single-event basis there is really no way good fields can be derived without having available high-quali ty, high-resolution, radar-rainfall intensity data, especially at convective conditions. If these models were to be used for determining the consequences of releases of hazardous material I believe that the establishment of a suitable network of weather radar would be an essential prerequisite. It seems something of a pity that no actual wet deposition data were available to see to what extent the model predictions (based on ensemble-average concepts) had any validity.

REFERENCES

Boris, J.P., and Book, D.L., (1973). Flux Corrected Transport I. SHASTA, an Algorithm That Works, J. Comp. Phys., 11, 38-69.

Briggs, G.A., (1971). Some recent analyses of plume rise observation. Procs. 2nd international clean air congress, 6 - 11 December 1970, Acad. Press (New York), 1029 - 1032.

Brost, R.A., and Wijngaard, J.C., (1978). A model study of the stable stratified planetary boundary, J. Atmos. Sci. 35, 1427 - 1440.

Bundesministerium des Innern, Bonn (1976). Räumliche Erfassung der Emissionen ausgewählter luftverunreinigender Stoffe aus Industrie, Haushalt und Verkehr in der Bundesrepublik Deutschland 1960 - 1980.

Businger, J.A., and Arya S.P.S., (1974). Height of the Mixed Layer in the Stably Stratified Planetary Boundary Layer. Advances in Geophysics, 18A, F.N. Frenkiel and R.E. Munn, eds. (Academic Press, New York).

Cats, G.J. (1980). Analysis of surface wind and its gradient in a mesoscale wind observation network. Mon. Wea. Rev., 108, 1100 - 1107.

CCMS study, (1981). Committee on the challenges of modern society, Report No. 126, Hout, K.D. van den, Dop, H. van, State of the art of inter-regional modelling.

CCMS study, (1981a). Committee on the challenges of modern society, Report No. 127, Aalst, R.M. van, Bergsma, F., Removal and transformation processes in the atmosphere with repect to SO_2 and NO_x.

Chem. Data Services, (1977, 1978). "Chemfacts Belgium" 1st Ed. 1977; "Chemfacts France" 1st Ed. 1978, London.

Deardorff, J.W., (1972). Numerical Investigation of Neutral and Unstable Planetary Boundary Layers, J. Atmos. Sci., 32, 1794 - 1807.

Dop, H. van, Haan, B.J. de, and Engeldal, C. (1982). The KNMI mesoscale air pollution model, KNMI Scientific Report, W.R. 82.6.

Dop, H. van, and Haan, B.J. de, (1983). Mesoscale Air Pollution Modelling, Atm. Environment 17, 1449-1456.

Endlich, R.M., (1967). An iterative method for altering the sinematic properties of windfields, Journ. Appl. Met., 6, 837-844.

Egmond, N.D. van, and H. Kesseboom, (1983). Mesoscale air pollution dispersion models; II Lagrangian Puff Model and comparison with Eulerian Grid Model, Atm. Env. 17, 2, 267.

Gear, C.W., (1971). Numerical initial value problems in ordinary differential equations. Prentice Hall, Inc., New Jersey, pp. 253.

Gland, H., (1982). Proceedings of the Intern. Techn. Meeting on Air Pollutions Modelling and its applications, Ile des Embiez, France.

Golder, D., (1972). Relation among stability parameters in the surface layer, Boundary Layer Met. 3, 47 - 58.

Gottlieb, D., and Orszag, A., (1977). Numerical analysis of spectral methods: theory and applications, Society for industrial and applied mathematics, Philadelphia.

Glückauf GmbH, (1982). Jahrbuch für Bergbau, Energie, Mineralöl und Chemie 81/82, Essen.

Haan, B.J. de, (1980). Procs. of the 11th ITM, NATO - CCMS, Amsterdam, November 1980. A comparison of finite difference schemes, describing the two-dimentional advection equation.

Herrmann, K., (1978). Zur Simulation interregionaler Transportvorgänge in der Atmosphäre, Ph. D. Thesis, TH-Darmstadt.

IEA, OECD, Energy Statistics 1976/1980, Paris 1982.

Killus, J.P., Meyer, J.P., Durran, D.R., Anderson, G.E., Jerskay, T.N., Whitten, O.f., and Reynolds, S.D., (1977), Continued Research in Mesoscale Air Pollution Simulation Modeling - Vol. V: Refinements in Numerical Analysis, Transport, Chemistry, and Pollutant Removal, SAI-report no. EF 77-142.

Luftreinhaltepläne, MAGS 1976-1982.

Metal Bull. Books Ltd., (1978), Iron and Steel Works of the World, 7th Ed, London.

Nieuwstadt, F.T.M., (1981a), The steady-state and resistance laws of the nocturnal boundary layer: theory compared with observations, Bound. Layer Met. 20, 3-17.

Nieuwstadt, F.T.M. (1981b), The nocturnal boundary layer, Thesis, Royal Netherlands Meteorological Institute (KNMI), De Bilt.

Reynolds, S.D., Meyer, J.P., Hecht, T.A., Whitney, D.C., Ames, J., Jocke, M.A., (1976). Continued research in mesoscale air pollution simulation modeling, Vol. II: Refinements in the treatment of chemistry meteorology, and numerical integration procedures, EPA-report 600/4-76-016b.

Reynolds, S.D., Tesche, T.W., and Reid, L.E., (1979), An Introduction to the SAI-Airshed Model and its Usage, SAI-Report no. EF 78-53R4-EF 79-31.

Richtmeyer, R.D., and Morton, K.D., (1967), Interscience Publishers, NewYork, pp. 405: Difference methods for critical value problems.

Sehmel, L.A., (1980), Particle and gas dry deposition: a review, Atm. Env., 9, 983-1011.

Tennekes, H., (1973), A model for the dynamics of the inversion above a convective boundary layer, J. Atmos. Sci. 30, 558-567.

Umweltbundesamt, (1980). Emissionsfaktoren für Luftverunreinigungen, Materialien 2/80, Berlin.

Veldt, C., (1982). Emission inventory for the mathematical modelling of photochemical air pollution in the mid-West of the Federal Republic of Germany, MT-TNO-report no. 82-014918, Dec.'82.

Veldt, C., (1983). Notitie Werkgroep Emissie-Analyse, MT-TNO, 5th April 1983.

Venkatram, A., (1980). "Estimating the Monin-Obukhov length in the stable boundary layer for dispersion calculations, Boundary Layer Met. 19, 481-485.

Wesely, M.L., and Hicks, B.B., (1977). Some factors that effect the deposition rates of sulphur dioxide and similar gases on vegetation, J. Air Poll. Contr. Ass. 27 (11), 1110-1116.

Wieringa, J., (1976). An objective exposure correction method for average wind speeds measured at a sheltered location, Anast. J. R. Met.Soc., 102, 241-253.

INDEX

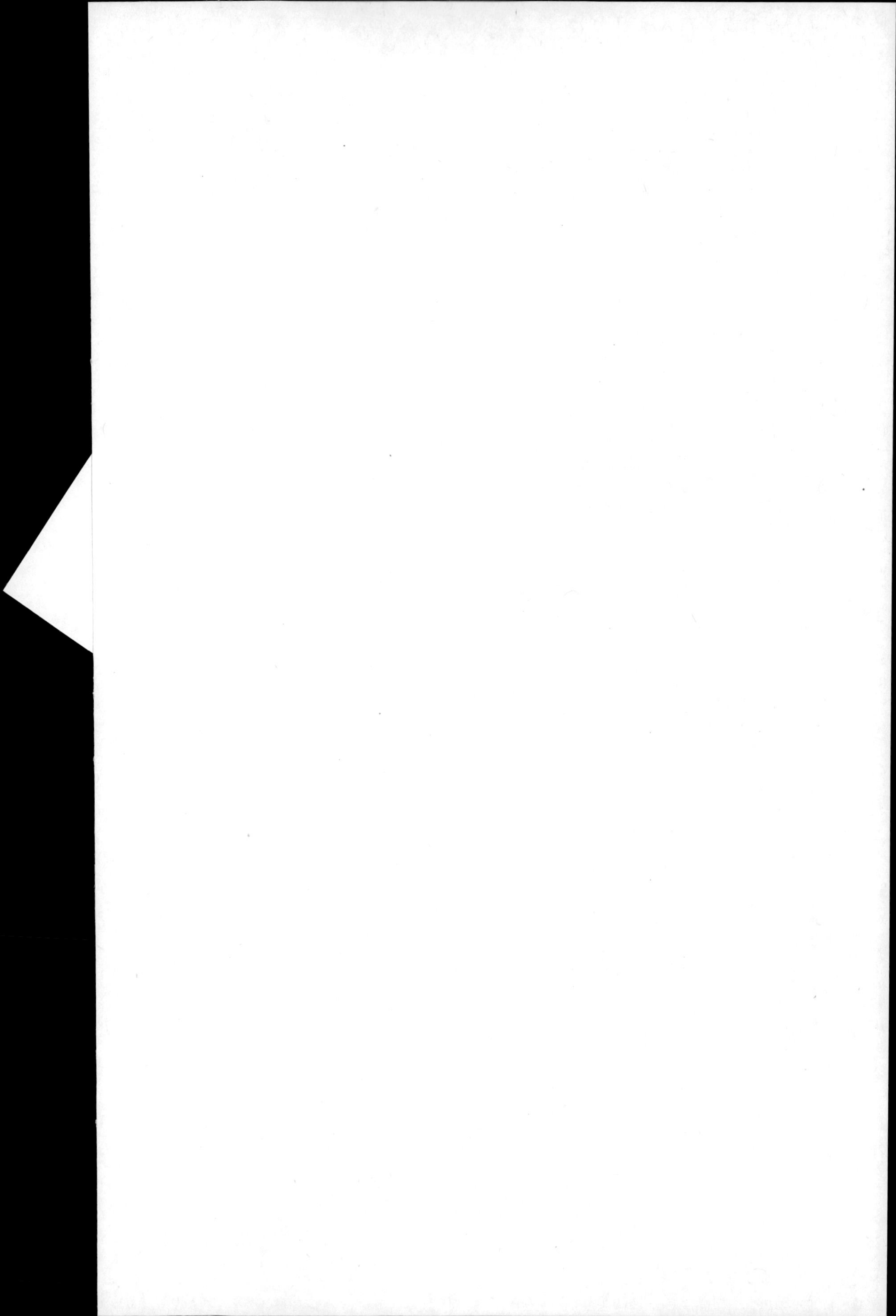

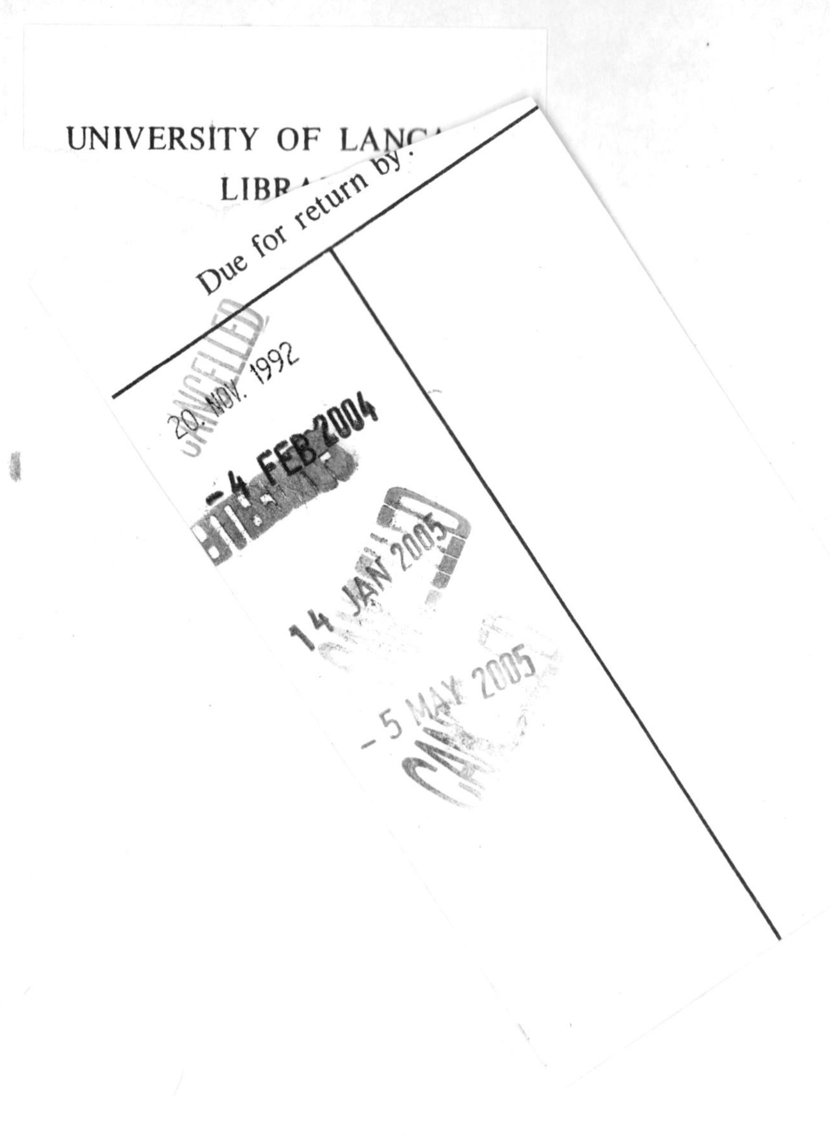